P9-DFL-732

'This is Adrian Berry at his best, partly because he writes so well and partly because he has a fascinating theme . . . I hope that this book will be republished in the year 2099, since by then we ought to have gained a much better idea of what lies ahead, and be able to gauge the extent to which his forecasts have been accurate. In any case, *The Giant Leap* is a great read, and I am confident that you will enjoy it immensely – I certainly did'

Patrick Moore, *Literary Review*

'What fun this book is, and how cheerful . . . I have not read anything as exciting since the editorials in *Amazing Tales* magazine, back in the Fifties, when the idea of human explosion into space stirred all of us'

Fay Weldon, *Daily Telegraph*

'Adrian Berry is an engagingly enthusiastic guide to the issues that we must address if we are to reach successfully for the stars. His patrician style, along with optimism and his gung-ho belief in scientific progress make for lively reading'

Graham Farmelo, *Sunday Telegraph*

'[I would praise] Berry's clear, good-humoured writing and the honest way in which he brings out the difficulties as well as the potentialities in human colonisation of space . . . I enjoyed reading this book'

John Michell, *Spectator*

'*The Giant Leap* is a book for both beach and bedside, providing an unlimited supply of pleasant dreams'

Reginald Turnill, former BBC Space Correspondent

Also by Adrian Berry

Non-fiction

The Next Ten Thousand Years
The Iron Sun
From Apes to Astronauts
The Super-Intelligent Machine
High Skies and Yellow Rain
Ice with your Evolution
Harrap's Book of Scientific Anecdotes
Eureka! A Book of Scientific Anecdotes
(a revised paperback version of the above)
The Next 500 Years
Galileo and the Dolphins

Fiction

The Fourth Reich*
The Empire in Arumac*
Koyama's Diamond
Labyrinth of Lies

Computer Software

Stars and Planets
The Wedgwood Benn Machine
Kings and Queens of England
Secret Key (with Keith Malcolm)

* Under the pseudonym of Martin Hale

The Giant Leap

Mankind Heads for the Stars

Adrian Berry

A TOM DOHERTY ASSOCIATES BOOK
New York

For my granddaughter Olivia

THE GIANT LEAP

First published in Great Britain in 1999 by Headline Book Publishing, a division of the Hodder Headline Group.

This book is printed on acid-free paper.

A Tor Book
Published by Tom Doherty Associates, LLC
175 Fifth Avenue
New York, NY 10010

www.tor.com

ISBN 0-312-87785-4

First U.S. Edition: September 2001

Printed in the United States of America

0 9 8 7 6 5 4 3 2 1

Adrian Berry was for nineteen years Science Correspondent of the *Daily Telegraph*, reporting on all branches of science until he retired in 1996 to devote more time to writing books. He still writes the monthly Sky at Night column for the *Telegraph*, as well as a regular column for *Astronomy Now* magazine. His columns in the former *Scope* page of the *Sunday Telegraph* were made into a book, *Galileo and the Dolphins*, published both in Britain and the USA.

He has written fifteen books, mostly of science and science fiction. *The Next 500 Years* and *Eureka! The Book of Scientific Anecdotes* have been published in eight countries and translated into six foreign languages. An earlier book, *The Next Ten Thousand Years*, sold half a million copies. He is a Fellow of the Royal Astronomical Society, the Royal Geographical Society and the British Interplanetary Society, and a member of the Planetary Society in the USA.

He is married with two children and lives in London.

Beyond lonely Pluto, dark and shadowless, lies the glittering realm of interstellar space, the silent ocean that rolls on and on, past stars and galaxies alike, to the ends of the universe. What do men know of this vast infinity, this shoreless ocean? Is it hostile, or friendly – or merely indifferent?

James Strong, *Flight to the Stars*

Who can doubt that this motion of the celestial vault, the eternal light of these lamps revolving so proudly above his head, the awful movements of this infinite sea, were established and are maintained for so many ages for his convenience and service?

Cicero, *De Natura Deorum*

For every action there is an equal and opposite reaction.

Isaac Newton's Third Law of Motion, which describes the propulsion of rockets and will make the project possible.

Contents

Acknowledgements

A large number of people have helped me in writing this book. I am especially grateful to my wife Marina, Dr Claes-Gustaf Nordquist, John Delin, and Dr Robin Catchpole, of the Royal Greenwich Observatory, for reading through the manuscript and making many useful suggestions.

However, despite the kind help of these people and those listed below, I must stress that the responsibility for any errors and misunderstandings is entirely my own. This applies also to the authors whose works are listed in the References and Notes, and in the Bibliography.

I especially thank Jovan Djordjevic for his splendid drawings which appear throughout the text.

Nor could I have written the book without the constant and kind encouragement of Lorraine Jerram, my editor at Headline, that company's copy editor Jane Selley, and my agent Rivers Scott.

I especially thank Keith Malcolm, who used his computing expertise to write the program that appears in Appendix III which partly solves the problem of starship navigation.

Grateful thanks are due also to Professor Frank Close, Professor Charles Pellegrino, Dr Patrick Moore, Dr Peter Morris, Lord Chetwode, Dr Roger Highfield, Sir Patrick Fairweather, Conraad Purchase, Robert Uhlig, Nigel McNair Scott, Dr Robert Zubrin, Harriet Cullen, John Walker, David Pryce-Jones, Mohamed Talib Al Salami, Gulshun Chunara, Chris Naylor, Jessica Berry, Jonathan Berry, Fabienne Pognant, Sir John Lucas-Tooth, Linda Kelly, Rachel Kelly, Rosanna Kelly, Christopher Rawlings, and Michael Tyrer.

I also had much useful assistance from the staff of the libraries of the Royal Astronomical Society and the Royal Geographical Society, and from the London Library, the Science Museum Library in London, and the South African Library in Cape Town.

Adrian Berry
AdrianBerry1@compuserve.com

Introduction

Man's quest to reach the planets of other stars began with a gruesome judicial murder.

It happened exactly four centuries before the first year of the new millennium. On a winter's morning in Rome in 1600, a group of hooded and black-robed men known, without intended irony, as the Company of Mercy and Pity bound a prisoner to a stake in the Campo dei Fiori, the Square of Flowers. For the last time they demanded that he recant his beliefs, and for the last time he refused. And so they burned him. Such was the fate at the hands of the Inquisition of the philosopher Giordano Bruno, whose only offence (apart from insulting his judges) was to proclaim his belief that other stars had planets.

It was the first time in scientific history that anyone had made such a suggestion.* Yet Bruno was not a scientist as we understand the term. He was not a mathematician, like Nicholas Copernicus who preceded him. Nor was he an observer of the sky, like Galileo who came after him. He was a visionary and a debater, whose only tool was his imagination. But his insight proved accurate. The planets he imagined to be circling the stars we now know to be real.

His terrible death made his ideas famous. But for his murder by

---

* It is true that the philosopher Lucretius in the first century BC, and the Chinese astronomer Teng Mu in the thirteenth century AD, both believed in the existence of 'other worlds'. But neither understood, as Bruno did, that a habitable world must have a sun and that the stars are other suns.

the state, it is doubtful whether posterity would ever have heard of him. But the bigotry of the Holy Office has preserved his reputation. It is in a small part because of him that we now dream of migrating to some of those alien worlds from whose surfaces the Sun would appear merely as a faint star.

His story is worth retelling. Half a century before Bruno's death, in 1543, Copernicus had infuriated both the Catholic and Protestant Churches by proposing that the Sun, not the Earth, is the centre of the solar system.

In writing his book *De Revolutionibus Orbium Coelestium* ('On the Revolutions of the Heavenly Bodies'), Copernicus made every effort to mitigate the anger he knew it was going to cause.[1] It was the work of 30 years, but he did not publish it until the last year of his life – receiving the first printed copy as he lay on his death-bed. He carefully dedicated it to the Pope, Paul III, addressing him as 'most holy Father', before proceeding to demolish, in remorseless mathematical detail, the Earth-centred system of 'epicycles' that had been official dogma for the previous 1,300 years.* A friend (without Copernicus's knowledge) even inserted a mollifying preface claiming that his proposal was purely hypothetical and not meant to be taken literally.

All these efforts to appease were in vain. Copernicus's system of astronomy was considered outrageous, both because it contradicted ancient wisdom, and because it dethroned Earth from its supposedly dominant position in the cosmos. Martin Luther called Copernicus 'anti-Biblical and intolerable', and *De Revolutionibus* was put on the Papal Index of forbidden books, where it remained until 1835!

Bruno, who followed in his footsteps, was less tactful. He saw religious dogma as particularly evil, and he did not hesitate to say so. In several books he set forth his ideas about good and evil. Wisdom, respect for reason, and the desire to discover truth were the only moral virtues. The enemies of these virtues were priests

---

* Its author was Claudius Ptolemy (AD 100–170), whose book the *Almagest* ('The Greatest') stated that the Sun, Moon and planets all went round the Earth in perfectly circular orbits. The stars, all at equal distance from us, did the same. Ptolemy was just as dogmatically wrong about geography. He proclaimed that no ship could cross the equator because it was so hot there that it would catch fire. It was not until 1497 that Vasco da Gama sailed round the Cape of Good Hope and proved that Ptolemy was a charlatan.

of all faiths, who were dissolute, avaricious, pedantic, and breeders of eternal dissensions and squabbles. Holy mysteries were rubbish, prayers were useless, and miracles were either tall stories or conjuring tricks.

But it was in scientific speculation that Bruno excelled. He took Copernicus's proposal still further, and maintained that not even the Sun was the centre of the universe. He was the first to realise, or at least the first to assert, that the stars were *other suns*. The only reason why they did not light up the sky as ours does, and that they appeared only as pinpoints of light, was that they were very far away.

And since innumerable suns existed, 'innumerable Earths must revolve around these suns in a manner similar to the way the seven [then known] planets revolve around our Sun. Living beings inhabit these worlds.'[2] Even though we do not yet know whether the part about living beings is true, it must have been a tremendous achievement to have reached this conclusion by imagination alone, without instruments and ignorant of mathematics. Passages from some of Bruno's books were so far-seeing that they might have been published in the twentieth century. An extract from one of them, *De La Causa* ('On the Causes of Things'), might have been written by Einstein:

> There is no absolute up or down, as Aristotle taught; no absolute position in space; but the position of a body is relative to that of other bodies. Everywhere there is incessant relative change in position throughout the universe, and the observer is always at the centre of things.[3]

In Bruno's early youth, he had briefly been a member of the Dominican order before he renounced religion in disgust. It was this that was to lead to his eventual martyrdom as a heretic and a lapsed monk. He spent many years wandering about Europe, writing, lecturing and debating. He had personal interviews with King Henry III of France and with Elizabeth I of England (who didn't like him, considering him radical and subversive). He met Shakespeare in a London printing shop, and in Oxford he held a debate with learned doctors about Copernicus's astronomy, at which, according to his own account, he was completely victorious. At last, in 1593, at the age of 45, he longed to return to his native Italy. A 'friend' offered him a house in Venice, but when he arrived in that city the friend denounced him to the Inquisition.[4] He spent the next seven years in prison, and was finally burned.* The Judges of the Holy Office condemned his views about other suns and other worlds, and sentenced him to death, in ponderous and repetitive language:

> We hereby, in these documents, publish, announce, pronounce, sentence, and declare the aforesaid Brother Giordano Bruno to be an impenitent and pertinacious heretic, and therefore to have incurred all the ecclesiastical censures and pains of the Holy Canon, the laws and the constitutions, both general and particular, imposed on such confessed, impenitent, pertinacious and obstinate heretics. We ordain and command that thou must be delivered to the Secular Court, that thou mayest be punished with the punishment deserved.
>
> Furthermore we condemn, we reprobate, and we prohibit all thine aforesaid and thy other books and writings as heretical and erroneous, containing many heresies and errors, and we ordain that all of them which have come or may in future come into the hands of the Holy Office shall be publicly destroyed and burned in the square of Saint Peter before the steps and that they shall be placed upon the Index of Forbidden Books, and as we have commanded, so shall it be done. Thus pronounce we, the undermentioned Cardinal General Inquisitors.[5]

To which Bruno made the spirited reply: 'Perchance you who pronounce my sentence are in greater fear than I who receive it.'†

This book is about Bruno's 'other suns' and their planets, and the plans that are being made to travel to them. Today, we not only know, as he could just speculate, that other stars have their attendant worlds, but we also dream of migrating to them.

---

* In 1889, as a result of a strong popular movement, a statue of Bruno was erected in the place of his execution, where it remains to this day.

† This was typical of Bruno's aggressive and defiant style of defence. He might have been pardoned if he had been more polite. As Isaac Asimov remarked, 'no one since Socrates worked so hard and with such determination to secure his own conviction'.

Several massive alien worlds have been found, of a size akin to the Sun's giant planet Jupiter, the only kind of planet we can detect with our present technology. But astronomers are also convinced that Bruno was right about the existence of 'innumerable Earths', planets the approximate size of our own that would be suitable for colonisation.

I call this book *The Giant Leap* after Neil Armstrong's prophetic statement when landing on the Moon in 1969: 'That's a small step for a man, one giant leap for mankind.' For interstellar flight will be the greatest of all leaps. It will be the ultimate achievement in space travel. Attaining it will be tremendously difficult, but not impossible. It is not a matter for the next fiscal year or for the next generation. It is unlikely that we will be able to do it even during the twenty-first century. For the distances between the stars are so huge, and the speeds needed to traverse them within reasonable voyage times so enormous, that the feat appears to us as difficult as crossing the Atlantic in a jet aircraft might have seemed to the people of the eighteenth century.

In short, nobody knows exactly how it will be done. I have therefore outlined several different schemes by engineers and physicists to accelerate spaceships to hundreds of millions of kilometres per hour, substantial percentages of the speed of light.

If indeed the speed of light really is an absolute speed limit. Perhaps one day it just *may* be possible to fly to the planet of a star hundreds of light years away for one's Christmas shopping and return in time for the festivities. The space agency NASA, as will be seen, is holding seminars to explore the idea.

Because all these schemes and ideas differ from one another, it has been impossible to write a 'clear-cut' book with a single thread. If at times, in some chapters, I appear to contradict what I said in others, this is inevitable in any discussion of such a futuristic project. If someone in the eighteenth century *had* set out to prophesy the coming of jet aircraft, he would certainly have had to describe some differing contemporary ideas about what they would look like and how they would fly.

He would also have had to answer the same objection that arises here, that the idea must seem wholly fantastic; not merely on technical and economic grounds, but on those of motivation.

Most writers about starships concentrate entirely on the technical problems. I try to go much further and explore some of the social and political aspects of the subject. What government, for example, would *willingly* assist a significant number of its people to vanish to places where they would never again have to pay any taxes? Why would any government wish to spend huge sums on interstellar flights at the risk of neglecting its domestic constituents? An obvious answer is that no government will want to do so, and that such missions will be privately funded. A less obvious one is that big government as we know it will largely disappear.

Why should anyone *want* to leave the solar system far behind and settle on other worlds, planning never to return? For people will indulge in interstellar travel not merely for scientific exploration (although that will play its part), but to *populate* the tens of millions of habitable worlds that are believed to exist in that vast cloud of hundreds of billions of suns that form our Milky Way galaxy. I attempt to show how they will be driven to this by the migratory urge that for billions of years has impelled man and his ancestors to roam, to explore.

Add to this the likelihood of profitable financial investments made possible by the shrinkage of time in vehicles travelling at very high speeds predicted by Einstein's special theory of relativity, and more practical motives become apparent. (After all, people invest in cocoa, wheat and other commodities, so why should they not invest in time?)

Any prediction that people will fill the galaxy seems to demand some discussion of every conceivable aspect of the subject.

How will it be possible to navigate accurately amidst the star-filled cosmos? What kind of ship will keep its crew sane for voyages that take years? What sort of 'inflight entertainment' will they require if they are not to go mad with boredom? Will not some people, instead of living on planets, prefer to 'make their own worlds' and inhabit comets? How long will it be before the Sun ceases to be benign and compels mankind to leave this solar system or perish? And is there not a need, comparatively speaking, for haste in preparation? For there will come a time when it will no longer be possible to travel to the stars because there will be no stars to travel to. The finale of the universe is one of desolation without end.

Finally, apart from the natural events mentioned in the previous paragraph, how can we be sure that all this is not science fiction? We cannot be absolutely sure, but it does seem reasonable to call it exceedingly probable. It merely follows a long-term trend in history to its ultimate conclusion. Soon after the revolution in science begun by Copernicus, Bruno and Galileo, there started a parallel revolution in *technology* that ever since has accelerated at a breathtaking rate – to the extent that today a new computer is said to be obsolete even before it is unpacked from its box. Consider the thousands of technical inventions we have made, most of them previously considered impossible by the most learned experts. If these, why not also starships?

This onrush of technical progress creates a tide in history, a tide that is virtually unstoppable because it is so massive, which will carry us to the stars. There is a discredited theory, known as the 'Great Man' view of history, that future events are as unpredictable as the weather, at the mercy of brilliant or idiosyncratic individuals. In the words of the historian Thomas Carlyle, 'Universal history, the history of what man has accomplished, is at bottom the History of the Great Men who have worked here.' But to Chancellor Bismarck, who knew a great deal more than Carlyle about the dark details of politics, events that have the strongest influence on the future were invariably made by those, like himself, who *just happened* to be the right person at the right time and the right place. As he put it: 'The statesman's task is to hear God's footsteps marching through history, and to try to catch on to His coattails as He marches past.'[6]

With these reflections in mind, let us look forward to the first manned interstellar flight, and see how its crew will fare.

Part One

QUEST FOR THE HEAVENS

Chapter 1

An Alien World

I have a dream that some day we'll build a telescope powerful enough to photograph the surface of an Earth-like planet orbiting a distant star, with enough resolution to distinguish clouds, continents, oceans.

Dan Goldin, administrator of NASA

Leave your home, O youth, and seek out alien shores. A wider range of life has been ordained for you.

Petronius, *Fragments*

The time is about two centuries from now. A vessel many tens of metres tall lands vertically in a clearing on an uninhabited world of forests, rivers and oceans. The jets from the exhausts beneath it thunder as it slows to zero speed, its tripod of legs made of exotic material taking the strain.[1] After a long delay, a dozen or so people cautiously descend from it. Hundreds of kilometres overhead circles its much bigger mother ship filled with people many times their number, who are slowly collecting their wits and their belongings after a voyage of several years. This will be the first time that people have landed on a world beyond the solar system, and they do not expect ever to return.*

---

* All except one. For certain reasons, it will be advisable that one person among the colonists *must* return to Earth after a starship has reached its destination planet. This is explained in Chapter 8, 'The Ascent of Rip van Winkle'.

They are all armed, these scouts who form the vanguard of the expedition; not because they know of any special danger, but because their training has taught them that alien planets are liable to be the domain of dangerous wild animals. They are sure there is no intelligent life in possession of advanced technology on this planet (that was one of the reasons for choosing it), because the radio emissions of such beings would have been detected from our solar system far away before the expedition set out. They know similarly, although they had come prepared to take the risk, that there are no primitive yet intelligent life forms, beings who, without technology, would be in the same condition as mankind for most of the history of *Homo sapiens* – because their existence would have been detected from orbit. Only one thing is sure: that lacking any such native competition, the planet is favourable for colonisation.

It is at about the right distance from its parent star to sustain life. It is in a 'Goldilocks orbit'; like the porridge in the fairy tale, it is neither too hot nor too cold.[2] It has approximately the same mass as Earth, which means that its gravity will be similar to what their ancestors experienced for countless ages. Its atmosphere, like Earth's, is a breathable mixture of oxygen and nitrogen with no detectable toxic gases. The number of active volcanoes is not excessive, nor are there fast-moving tectonic plates which could have produced an unbearable number of earthquakes. This planet does not rotate too fast, which would produce violent weather conditions; nor does it rotate too slowly, a condition that would create extremes of temperature. As a consequence, wind speeds are moderate, which is fortunate since, if they persistently exceeded terrestrial gale force, blowing at more than 100 k.p.h., living conditions on this new world would be intolerable.

I mentioned oceans. Without large bodies of water, there will be little possibility that the colonists could have easy access to rare and precious metals with which to build up industry. These are copper, silver, gold, uranium, thorium, zinc, lead, tantalum, niobium, tungsten and molybdenum; the 'strategic metals' whose prices are quoted daily in newspapers. As one scholar explains:

> Without liquid water there would be no leaching of lead, zinc and copper from basalt rock. Neither would there be any rain to wash tin, gold, uranium-bearing minerals, thorium minerals

> and ores of niobium, tantalum, molybdenum, tungsten and the rare earths from granite rocks and concentrate them.[3]

Fortunately there is every sign that planets do have oceans whenever their compositions enable them to. Mars had at least two oceans several hundred metres deep billions of years ago when it was a much warmer world than it is today.[4] And the entire galaxy appears to be filled with water in the form of ice. Not only is ice the main component of comets; it is also present in vast quantities in the interstellar dust clouds from which stars and planets are formed.[5]

It is impossible, of course, that this hypothetical world could be *exactly* like an uninhabited Earth. No two planets, even in the entire universe, could be absolutely identical. Even if there were tens of trillions of Earth-sized planets, that would still defy probability. Yet there are many ways, some of them quite surprising, as Stephen Gillett points out in his book *World-Building*, in which a planet could be composed or situated and still be habitable for people wearing ordinary clothes and without breathing apparatus.

To be in a Goldilocks orbit, the planet would not even need to circle a star directly, as ours does, and take all its warmth from that star. It might be a giant satellite of a 'brown dwarf' star that in turn circled a normal star.* (Such a world is the planet Haven in Jerry Pournelle's highly convincing *Warworld* science fiction stories.) It would obtain some of its warmth and light from the dwarf and some from the star, greatly extending the size of the 'ecosphere', the region round a star where habitable planets could exist. In such a case, says Gillett, the dwarf might appear in the skies of their world 20 times wider than the Sun appears in ours, 'a vast ball perhaps glowing deep, dull red, with streamers and wisps of darker clouds and possibly storms and belts like Jupiters'.[6]

The sun of this imagined world is a very ordinary star, similar in age and mass to our own. If it were not, the settlers would not

* Brown dwarfs could be described as 'almost stars'. Believed to be among the commonest objects in the universe, they are too massive to be planets and too small to be true stars. They give out heat from the gravitational contraction of their mass, but not enough heat to let them undergo nuclear fusion and shine as a true star. To be a brown dwarf, it is generally considered that an object must have at least six times the mass of Jupiter and about 1 per cent of the Sun's mass.

have considered coming here. But there is a good chance of there being one big difference. Like about half the stars in our Milky Way galaxy, and unlike the Sun, it is likely to have a companion star.[7] If the two stars were close together, say the approximate distance between the Earth and the Sun, the planet's orbit and the regularity of solar radiation reaching its surface might be unbearably erratic. But this second star, to assume a common enough situation in stellar astronomy, may well be four billion kilometres from its companion, about the distance of the Sun from Pluto, normally the Sun's most distant world, and so in the night skies it will shine only as an unusually bright star. It has little or no effect on the planet's orbit, and the effect, from the point of view of any radiation and gravitational disturbance, is as if it did not exist.[8]

The same could well apply to triple and multiple star systems, of which there are many in the galaxy. One such is Epsilon Lyrae, the famous 'double double', where *two pairs of stars* orbit each other. The two components of this star system are so far apart that they can even be distinguished with the naked eye. And the two stars of *each pair* appear to be far enough apart for either to have a habitable planet.*

But to return to this imagined world. It has one greater asset, the main reason why it was chosen as a destination. It has abundant plant life which can form the basis of large-scale agriculture. The would-be colonists number only about a thousand, roughly the size of a small isolated country town in Australia, and comfortably above the number which might cause their descendants to suffer from hereditary disease as a result of in-breeding.[9] But this population will grow. The colonists have found what will eventually be a home for tens of millions of their descendants.

Such a scene is reminiscent of many a science fiction novel written in the last four decades, and science fiction is, at its best, the one form of literature that deals intelligently with such

---

* Amazingly, the 'double double' has an exact copy elsewhere in the same constellation. Between the stars Gamma and Theta Lyrae lies the 'double double's double', whose configuration of four stars appears to be identical. Perhaps there is some astrophysical reason for such a pattern, and perhaps the 'double double' has more doubles of itself elsewhere.

scenes as these.* But there is a danger in this scenario. An initial population of a thousand, probably the maximum that is practical for the pioneering expedition of a starship, *may not be nearly enough* to set up a viable colony on their planet of destination. As the economist William Hodges puts it: 'The first generation may regret, and subsequent generations bitterly resent the fact that they ever left the solar system. The political organisation is likely to become authoritarian, and the leader may resort to corporal punishment.'[10]

Their situation is even more precarious than that which faced the colonists who settled the remotest shores of Earth. They at least could call for fresh provisions from home. But the first explorers of the planet of another star will be as cut off by distance as if their home world did not exist.

A shortage of tools and equipment could bring about the crisis that Hodges fears. Everything the colonists need they must either bring with them or make. A lack of adequate equipment will make it impossible for them to build up the essential infrastructure of homes, factories, greenhouses, machine shops, foundries, electricity-generating plants, telescopes, and countless other installations. As Hodges remarks, 'water, metals, oxygen, carbon and nitrogen will be available in effectively infinite quantities, but not in an immediately useable form'.[11] Even to make the things that are absolutely necessary to life, they will need a huge variety of tools and machines.† Without these, their way of life will soon become medieval or even more primitive. It would be ironic indeed if, after a voyage that demanded a level of technology far beyond ours, the state of the colony should have degraded, within a few generations, to the level described in a study of Europe's Dark Ages:

> Their villages were frequently unnamed. If war took a man even a short distance from a nameless hamlet, the chances of his returning to it were slight; he could not identify it, and finding his way back was virtually impossible. Each hamlet

---

* As Arthur C. Clarke has remarked, 'I would claim that *only* readers of writers of science fiction are really competent to discuss the possibilities of the future.'

† Perhaps one of the most important artefacts, Hodges suggests, will be a machine that can make microprocessors from recycled garbage.

> was inbred, isolated, unaware of the world beyond the most familiar local landmark: a creek, or mill, or tall tree scarred by lightning. Generations succeeded one another in a meaningless, timeless blur.[12]

Countless expeditions and settlements, even the best organised, have in the past ended in decline or extinction. The first English colony in North America, consisting of 120 people who settled on Roanoke Island off North Carolina in 1587, vanished. Three years later, no trace of any habitation could be found there.[13] Captain James Knight disappeared with his crew of 40 men in 1719 while stranded in Hudson Bay during an attempt to find the North-West Passage. Cut off by ice and apparently driven mad by mosquito bites, they either killed themselves or fled.[14] Sir John Franklin and all his crew died in a similar Arctic search in 1847, perhaps from carbon monoxide poisoning from their own stoves.[15] Tristan da Cunha island in the South Atlantic was first settled in 1817 by three people, a population that has since increased to more than three hundred. But they suffer from hereditary deafness, a condition almost certainly caused by incestuous marriages.[16]

Perhaps the closest parallel to the alien planetary settlement I have described could be the ominous story of the Polynesian settlers of Pitcairn, Henderson and Mangareva islands in the tropical Pacific. This episode is particularly apt because it concerns not one island but three; and the planetary colonists, having settled on one world, will certainly try to colonise another nearby. (This will be easy for them to do if they are in a double or multi-star system.) The anthropologist Jared Diamond tells the story of what happened in the Pacific.[17]

When the mutineers from Captain Bligh's *Bounty* reached Pitcairn Island in 1790, fleeing from the vengeful British Navy, they found that others had lived there before them.* There were temple platforms, rock carvings and stone tools. A much later archaeological investigation discovered that for several centuries

* Their refuge was not discovered until 1808 because Pitcairn Island, although known since 1767, was inaccurately placed on Admiralty charts. By that time the British Navy was too busy in its struggle against Napoleon to mount a fresh expedition to capture the mutineers.

from about AD 1000, Pitcairn, an island of 27 square kilometres, Mangareva, about the same size but with a huge lagoon 25 kilometres wide, and Henderson, 36 square kilometres in area, had been a united civilisation of the Polynesians. Thousands of kilometres from other major islands and thus effectively isolated, the three islands were several hundred kilometres apart from each other, but that was nothing to the fast canoes of the Polynesians, who sailed between them in a few days, prosperously bartering goods.

The Polynesians were the most successful colonists in history. In the space of two millennia they created settlements on tens of thousands of islands that spanned the oceanic world. But this civilisation on these three isolated islands was one of their least fortunate ventures. By 1500 it had collapsed, and the populations of Pitcairn and Henderson had vanished. As Diamond explains:

> Immigrants came to a fertile land blessed with apparently inexhaustible natural resources. It lacked a few raw materials that were important for industry, but they were readily obtained by overseas trade. For a time, the land and its neighbours prospered, and their people multiplied.
>
> But the population of the rich land grew too large for even its abundant resources. As its forests were stripped, and its soils eroded, the land could no longer nourish even its rich population, let alone grow food for export. Then, as trade declined, the imported raw materials began to run short. A kaleidoscopically changing succession of local military leaders overthrew established political institutions, and civil war spread. To survive, the starving populace turned to cannibalism. The fate of their former overseas trading partners was even worse: deprived of the imports on which they had depended, they ravaged their own environments until no one was left alive.[18]

The worst mistake the Polynesians made in this case was their self-indulgence. They cut down most of the forests to plant gardens. Rain carried off the denuded topsoil, and soon there was hardly any land for cultivating crops and no more tree trunks with which to build canoes. Inter-island trade faltered and stopped. People started to take by force what they could no longer get by barter. A life of luxury declined into one of poverty and barbarism.

There is only one way that the planetary colonists can avoid making these blunders. They will have to decide, from the beginning, to live in a comparatively primitive fashion, in equilibrium with their environment, and not, like the Polynesians with their three islands, imagine that they have reached a land of boundless riches. The planet and its neighbours may *possess* boundless natural riches, but the colonists will at first be far from possessing the tools to exploit them. On disembarking from their starship they will have to move from the technology of the twenty-second century back almost to the nineteenth.

Their descendants will enjoy much higher levels of technology, but for the first generation, and perhaps the second, there will be little but the hardest labour. Perhaps Peter F. Hamilton, in his 1997 novel *The Reality Dysfunction*, with his colonists in their wooden huts and uncomfortable boats, has best captured the sheer ruggedness of what it will be like to live as the first generation on an alien world without hope of assistance from Earth.[19]

There will be no roads, and so there will be no use for cars, trucks, or petrol. There would have been no room in the starship for such cumbersome luxuries, and even if there had been, there would have been no means of repairing them when they broke down. Instead, the colonists are likely to use horses and donkeys, and oar- and sail-driven boats, for transport. (A starship will need to carry such animals, perhaps in frozen embryonic form to save space.) Their earliest sources of energy will be windmills or hydroelectric turbines. They will have brought with them a nuclear fission reactor, but it will take some time to assemble and test it, and until then wind or water power will have to suffice. Fission will be far easier to handle than nuclear *fusion*, since a fission reactor presents far fewer difficulties in construction and repair.

There will be innumerable works of construction. The most urgent of all will be for individual needs. For example, the shoes of the colonists will wear out, and they will need a cobbler. But the cobbler will eventually die, and so he will need an apprentice. (They *could*, of course, have brought with them a vast supply of spare shoes in the ship, but doing so would have taken up a prohibitive amount of valuable space. The same objection applies to the transport and storage of supplies of all other spares.)

It will be imagined, at a first consideration, that the colonists must include an enormous number of specialists.

A vet, it will be supposed, will be needed to look after the countless animals they have brought with them, and the vet also will need an apprentice. There must also be mechanics, mining engineers, electronic experts, people responsible for food production and at least one doctor and a dentist (all with their apprentices). Some writers have suggested that these difficulties could be vastly simplified by having one single 'medical person', who could combine the roles of pharmaceutical expert, chemist, vet, dentist, doctor, and all-purpose surgeon.

But this is absurd. No single person could possess such vast arrays of skills and apply them to a rapidly expanding population. Besides, the proportion of 'experts' responsible for the welfare of this population, whether working full-time or permanently on call, would have to be so great that an insufficient number of people could be spared to pursue their private aspirations, and the colony would become heavily bureaucratic. While growing in numbers, it could never grow in achievement. The only solution will be to have as many experts as possible, not in human form, but in that of computer software programs.

These kind of programs are called 'expert systems'.

In several important ways, they are superior to human experts. A human dentist or doctor – or a specialist in some branch of dentistry or medicine – may be a desirable person to have on a distant planet, but it is very expensive to feed him and to get him there. A computer DVD-ROM disk (short for digital video disk, and the successor to CD-ROMS) on the other hand requires neither food nor heating nor clothing, it occupies no more space than a postal envelope, and it can be kept on a shelf until needed. Moreover, it contains about four billion bytes of data, about as much information as a full-length feature film, and nearly 60,000 times more than a book of average length.[20] By the time starships are built, such disks will no doubt carry thousands of times more information even than this.

An expert system, as its name suggests, gives expert advice when told that a problem exists. A 'doctor expert' will ask the patient questions about his symptoms and then reach a conclusion, either tentative or certain depending on the patient's answers. Most important, when challenged, *it will explain why it is asking any given question.* It will then not only give its

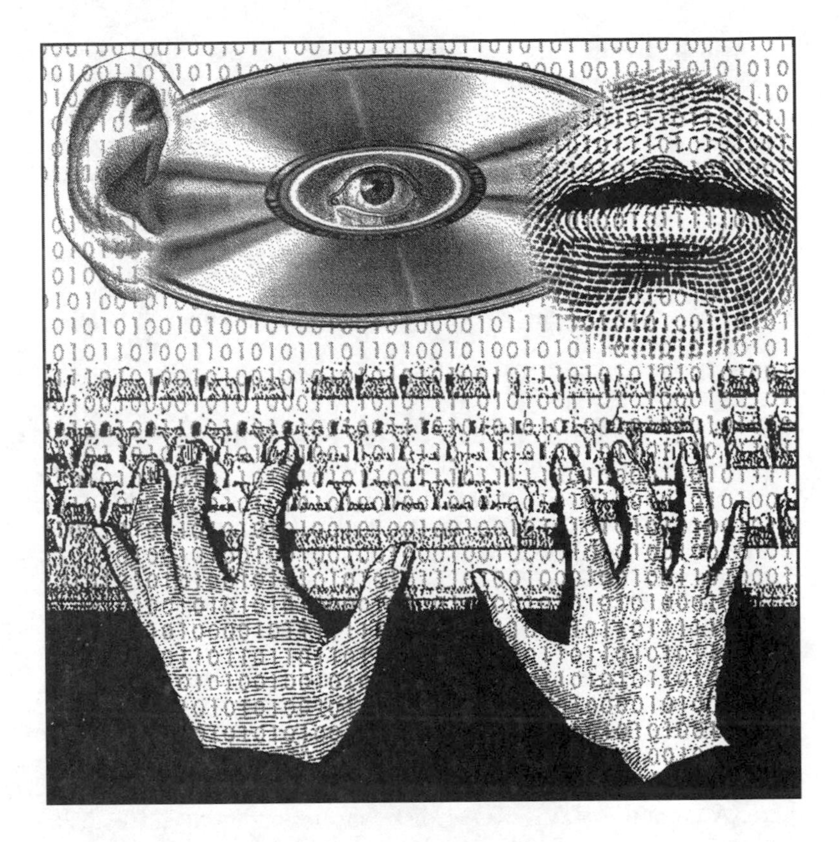

diagnosis, but will also indicate how it reached that diagnosis and indicate just how certain it is of it. It may then also go on to advise what should be done about it.

One means of creating such programs has been developed by the British computer expert Chris Naylor, author of the book *Build Your Own Expert System*. It is a software package called *XMaster* and is an 'Expert System Shell'. This is a program which makes it very easy to produce expert systems in specific fields such as medical diagnosis. The idea is that 'knowledge bases' of specific expertise can be created to solve any problem which is capable of a factual solution. It can then be loaded into *XMaster* for consultation at any later time.[21]

I tried out Naylor's *XMaster* using a demonstration medical knowledge base. I asked it if it was likely that I had influenza. Here is the question-and-answer session that followed:

XMaster: Do you have a high temperature (over 38° say)?
Me: Yes.
XMaster: Do you feel generally ill?
Me: Yes.
XMaster: Do you have symptoms which have been present for some time – possibly six weeks or more?
Me: No.
XMaster: Are you sneezing a lot?
Me: Yes.
XMaster: Do you have a sore throat?
Me: Yes.
XMaster: Do your muscles ache?
Me: Yes.

On the strength of these answers, *XMaster* was nearly certain that I was suffering from influenza.* (When I reran the consultation session, answering all the questions in the same way except that I stated that the symptoms *had* lasted for six weeks or more, it then ruled out the possibility of influenza, since its symptoms seldom, if ever, last for six weeks.)

As for surgery, when it is needed, doctors could be guided by video records of similar operations performed on Earth. But the key to keeping the community healthy, say two experts on survival on an alien planet, will always be access to information.[22]

Similarly, the colonists will need other expert systems to solve all manner of problems they will face. Agriculture, nuclear engineering, geology, oil and coal exploration, astrophysics, seismicity, environmental chemistry, meteorology, the identification of unknown substances – they will need expertise on these subjects and many more, and it would have added greatly to the

---

* It can be justly argued that such software does not remove the need for physical testing. But a medical testing facility does not have to be *specialist*. A single device or machine can test for a huge variety of diseases. One or two such test facilities will suffice for an entire community. And such a laboratory takes up far less space than a dozen specialist doctors.

costs of the expedition if they had been forced to bring along humans who had specialised expertise in each of them.[23]

Later generations on the planet, if all these difficulties have been overcome, may look back on the early stages as a heroic saga – and justifiably, since those who took part in them will have been courageous and determined people. But now their standard of living, their population, and their technological abilities will have risen rapidly. They will be back to the twenty-second century and beyond. They will have turned their world into something more than Earth, and without Earth's environmental problems. They will have founded sub-colonies on other planets in their own stellar system, since it is likely to consist of two or more stars. And as the centuries pass by, they will send colonies to the planets of more distant stars, and so on.

The darker possibility is that the early difficulties will *not* be overcome, and that the colonists will perish like the Polynesians in their three isolated islands, their pathetic traces discovered by interstellar archaeologists ages later.*

If this occurs, the most probable cause will be that one or more influential member of the first generation colonists will have been a person of flawed character.

This raises an all-important problem: how should people be chosen for a starship mission? The most sensible approach may be to turn the question around and ask: on what grounds should aspiring candidates be rejected?

Obviously, the personalities of applicants will be examined with the utmost rigour to ensure that they are the kind of people they purport to be. Far more rigour indeed than that faced by passengers boarding an airliner who are usually only prevented from doing so if they have no ticket, appear to be drunk, or are carrying a concealed weapon. It will not be enough to disbar only those who are lazy, stupid, quarrelsome, generally obnoxious,

---

* Diamond tells the story, which may be apposite, of the Indian colony on San Nicolas Island, off Los Angeles, whose population dwindled until it was reduced to a single woman who survived in complete isolation for 18 years. He adds: 'Did the last Henderson Islanders spend time on the beaches, generation after generation, staring out to sea in the hopes of sighting the canoes that had stopped coming, until even the memory of what a canoe looked like grew dim?' For canoes, read starships.

suffer from diseases, hereditary or otherwise, or are incapable of sexual reproduction. A much deeper analysis of character and past history will be needed. This is not as difficult as it may sound. Even today, using the Internet, it is possible to gather an enormous amount of information about any given person. Within a very short time one can discover if someone has a criminal record, unpaid debts, or a history of mental disorder.

But that is the easy part. The problem is *impersonation*. There has been many a thriller whose turning point is that seemingly respectable Mr So-and-so, carrying all the right credentials, is not in fact Mr So-and-so at all but an imposter who has waylaid and murdered the original. The greatest danger is perhaps not of ordinary crime, for which the opportunities will be sparse, but of terrorism by a psychopath, either in the starship or on the planet of destination. The colonists will hardly want to be in the position of the house-guests trapped on an isolated island in Agatha Christie's novel *Ten Little Indians*, one of whom is a killer who sets out to slaughter the rest.

So much for the worst possibilities – apart from those I cover in later chapters. Whether the colonists ultimately survive and flourish will be up to them, their recruitment officers, and their ability to deal with conditions they have no means of foreseeing.

Chapter 2
Starships and Politicians

And were it not that they are loath to lay out money on a rope, they would hang themselves forthwith to save charges.
Richard Burton, *Anatomy of Melancholy*

So how do we get there from here?

Or perhaps more important, do we ever get there from here? Until recently, most astronomers at least have taken the view that, because of the huge distances, little-understood technology and gigantic costs, manned interstellar flight will never take place because it is close to impossible. As the astronomer Edward Purcell remarked in 1963, 'all this stuff about travelling round the universe in space suits, except for *local* exploration [i.e., to the Sun's other planets], belongs back where it came from, on the cereal box'.[1]

But more recently there has been a change of feeling, led not by astronomers, who tend to despair at the obvious and tremendous difficulties, but by engineers who see the prospect as an interesting challenge. The first author to predict starships had been the physicist J.D. Bernal in his 1929 book *The World, the Flesh and the Devil.* Astronauts, he suggested, would burrow into comets and convert them into starships, using their fiery tails that always point away from the Sun to repel oncoming matter that might otherwise dangerously collide with them when they reached high speeds. How these 'ships' would fare when their tails (consisting of matter vaporised and blown off by the Sun's heat) disappeared as they left the Sun's neighbourhood, he did not say. Yet his vagueness was more than made up for by his enthusiasm:

> Once acclimatised to space living, it is unlikely that man will stop until he has roamed over and colonised the universe, or that even this will be the end. Man will not ultimately be content to be parasitic on the stars but will invade them and organise them for his own purposes.[2]

The first prophet of interstellar travel to go seriously into practical details in a full-length book was a distinguished member of the British Interplanetary Society, James Strong, in his 1965 book, *Flight to the Stars: An Enquiry into the Feasibility of Interstellar Flight.** Although now largely out of date, Strong's book is impressive in its discussions of methods of propulsion and the almost intractable problems of navigation amidst what he called the 'shoreless ocean' of interstellar space. Then, in the sixties, came the television series which has endured to this day, *Star Trek*, with its Klingons and its 'warp speeds' and its 'Beam me up, Scotty', and other scientific improbabilities.† But *Star Trek* drew attention to the *possibility* of interstellar travel and aroused widespread interest in the subject. It was a few years after the birth of the series that the British Interplanetary Society published a technical report, 'Project Daedalus', showing that an unmanned space probe propelled by a thermonuclear rocket, which was in principle available at the time, could attain 12 per cent of the speed of light and reach Barnard's Star, six light years from Earth, within 50 years of its departure.

The flowering of electronics (which Strong did not mention, since computers barely existed in his time), progress in particle physics and growing knowledge of the environment between the stars led Eugene Mallove and Gregory Matloff (respectively an

---

* In Appendix I, there is a timetable of the important dates in the history of preparation for interstellar flight.

† 'Beam me up [or down], Scotty' would be a most environmentally unfriendly command. According to Lawrence Krauss, in his book *The Physics of Star Trek*, 10^{28} atoms, the total mass of a person's body, would have to be converted into energy for teleportation. This action would release the energy equivalent of 1,000 one-megaton hydrogen bombs! It seems that the only safe way to get the crew down to a planet's surface is in a landing vessel.

engineer and a mathematician) to publish in 1989 their highly modern *The Starflight Handbook*, which they called 'a working guide for the would-be star traveller'.

So much for words. Have any spacecraft actually been sent to the stars, as opposed to the planets? The surprising answer is yes. Four such unmanned craft have been sent. *Pioneer 10* and *Pioneer 11*, launched by NASA in 1972 and 1973 respectively, photographed Jupiter and then headed out of the solar system into interstellar space. The first of these carried a plaque designed by Carl Sagan, his colleague Frank Drake and Sagan's wife Linda, for the benefit of any aliens the craft might encounter. It was intended to show them how much we knew about the universe and what they should know about us. It gave the locations of the then 14 known pulsar stars in relation to the position of the Sun, Earth's position round the Sun, the frequency of the hydrogen atom, and an engraving of a naked man and woman, showing their sexual organs.[3]

Four years later, *Voyager 1* and *Voyager 2* were launched, also on trajectories that took them into interstellar space. More technically advanced than the *Pioneers*, they photographed Jupiter, Saturn, Uranus and Neptune before leaving the solar system. Officials, mindful of the furious protests from puritanical people

at the sexual illustrations on *Pioneer 10*, fitted them only with long-playing records called *Sounds of Earth* for the enlightenment of aliens.

But these craft were not aimed at any particular star. Nor did they travel at a speed that would enable them to reach the stars within reasonable voyage times. After being accelerated by planetary 'swing-bys' of Jupiter – as shown in fiction in the film of Arthur C. Clarke's novel *2010: Odyssey Two*, they finally left the solar system at speeds of a few tens of thousands of kilometres per hour, enabling them to reach the nearest star systems beyond ours within tens of thousands of years.*

The crew of a *manned* starship would clearly not be content with such aimlessness and low speeds. They would need a vastly swifter vehicle and a carefully planned destination. Moreover, the would-be star traveller has one overwhelming problem which I attempt to address in this chapter and the next: who is going to pay for his journey from which he plans never to return, and is the financing of the journey even possible in the present-day form of society with its excessive degree of government?

There is a general assumption among people who write about space that manned travel to the stars will take place as soon as such journeys become technically possible. First, it is said, after all the Sun's planets have been explored by unmanned probes (a process which is nearly complete), manned exploration of the planets will take place. This will be followed in turn by permanent settlements on these worlds and in some cases by colonisation. Everything will be managed in an orderly fashion by governments, either in the form of a World Space Council or by national space organisations like NASA or the European Space Agency. In the long term, so the argument goes, because the growth of wealth will roughly keep pace with the level of technological development, governments will be able to afford to finance these vast projects without neglecting their domestic constituents.

It is a charming scenario, but unfortunately, like many such idealistic visions of the future, it will not happen like this. The

* Spacecraft headed for the outer solar system often gain acceleration from a 'swing-by' of another planet. By briefly orbiting the planet, they 'steal' some of the energy of its rotation.

principal reason is that politicians tend to be uninterested in funding any project that takes more than four years to come to fruition – the time until the next election. And the design, building and testing of a starship will take about 50 years to complete. As the space expert John S. Lewis points out:

> Most politicians [in the United States] see basic research from the perspective of industry, which has become increasingly obsessed with the current quarter's balance sheet. Many industries see that basic research is a major expense this quarter, and that this year's basic research will not result in a single new product this year. They therefore have chosen to 'improve' their ledgers by cutting funding for basic research.
>
> That such behaviour is suicidal in the long run is not at all obvious to them, nor does it address their short-term needs . . . The most astute critic might see the single cause at the root of these difficulties: the ascension into top management of a generation of managers whose education trains them to count beans but leaves them perfectly ignorant of where future crops of beans come from.[4]

Politicians and bureaucrats similarly, if asked to support long-term technological projects, are liable to give the same negative reaction. In the words of another critic, 'they ask themselves: "Will this work? When, and how much will it cost?" and (if not aloud!) "*What will it do to my chances of re-election or promotion* (a) if it succeeds, and (b) if it fails?" '[5] Of course, since the project is so long-term, the administrator in question will probably never know whether it succeeds or fails. But he will still fear it and favour its cancellation. This is not only because news of its success or failure will come long after the end of his career and therefore cannot damage it or improve it; it is because administrators (I use the term to embrace both politicians and civil servants) tend to fear any project which threatens to introduce new science to the world.

New science and new technology can be, from an administrator's point of view, uncontrollable. At worst they can take away power from government and put it into the hands of the people. As a result of this, governments tend to hate and fear inventors and the coming of new technological ideas. The most obvious example of this is the Internet which tends to abolish international frontiers and enables individual citizens to communicate in absolute secrecy (of which much later). But there have been many others.

One of these had a profound effect on the nature of television. Back in the seventies there were only a limited number of television channels (it was long before cable and satellite TV), and a huge amount of television time was taken up with political talk shows, or 'talking heads' programmes, as they have been called, in which politicians endlessly argued and postured. Millions of viewers were captive to these dreary shows. Then came an epochal invention by the engineer Shizuo Takano, managing director of the Japanese Victor Company (JVC), who believed that the television set was a hopelessly underdeveloped machine which could be turned into an 'entertainment computer'. Viewers could watch what *they* wanted rather than what the programmers wanted them to watch. The result was his invention of the video recorder, which enabled people to watch pre-recorded films.[6]

It was a measure of the astuteness of at least one politician, President Mitterrand of France, that he foresaw the danger of this threat to his profession and for many months held up the import into his country of the new video machines. But such obstruction was in vain, and today there are hundreds of millions of these machines, and video rental shops in almost every urban neighbourhood.*

The VCR literally took the politicians off our screens, and they have always resented it. In 1998 in Britain, television chiefs informed party leaders that they were going to cut the number of political talk shows because 90 per cent of viewers found them boring, did not believe that politicians answered questions truthfully, and were baffled by political jargon like 'backbencher', 'select committee', and 'white paper'. Ministers and Members of Parliament protested furiously, probably in vain, and one must

---

* Takano faced obstruction, as inventors often do. His idea was challenged by government broadcasting authorities in the West, who thought that television had reached, and ought to have reached, its technological limits with the advent of colour. In Britain in 1977 the Independent Broadcasting Authority stated that it could see 'no future' in VCRs, which 'would appeal only to the minority who are choosy in their viewing attitudes and who can afford the equipment that makes it possible to be choosy'. The snobbery implied by the word 'choosy' is incredible.

confess that in contemplating their fury the only emotion that springs up is one of contemptuous amusement.[7]

Notwithstanding the lower esteem in which politicians are held, and the diminished interest which people take in their daily business, they have shown a tendency in recent decades to entrench themselves. They have responded to ever-increasing wealth with massive increases in taxation and, in the European Union, oppressive regulations imposed by non-elected officials. We appear to be governed, as always, by a permanent class of administrators who are determined to yield none of their prerogatives. Back in 1976, the science fiction writer Frank Herbert, the author of *Dune*, saw this as an immutable feature of history:

> Governments, if they endure, always tend increasingly towards aristocratic forms. No government in history has been known to evade this pattern. And as the aristocracy develops, government tends more and more to act exclusively in the interests of the ruling class – whether that class be hereditary royalty, oligarchs of financial empires, or entrenched bureaucracy.[8]

What, against this background, are the prospects that governments will finance or encourage the building of starships? It must be admitted frankly, practically none at all. There does not seem to be the slightest chance of their willingly permitting the development of a technology that threatens to allow people to escape from their jurisdiction for ever. Indeed, given Herbert's analysis, from their point of view they would be fools to do so. Starship construction is perhaps the ultimate technology of which man is capable, and it is safe to say that advances in almost *all branches of science*, whether in engineering, computing, astronomy, physics, mathematics, life science, medicine, propulsion, or robotics, tend to bring nearer the day of the starship.[9] Some far-sighted minds in government dimly realise this, or at least that there are levels in technology which, if ordinary people ever get their hands on them, will bring about situations that the state cannot hope to control.* The wresting of control of the

---

* There is evidence that, ever since the time of Copernicus and Galileo, politicians have hated the universe and wished it would go away. In 1825, US President

television set from the authorities will be as nothing to what may come! The response of the state, therefore, is to 'put on the brakes' as sharply as possible; to restrict the funding of all science that they oversee, and to hope that private industry will continue to restrict pure research in the way that John Lewis describes. As one expert summarises the results of this timorous policy:

> The trend with government-controlled projects has always been towards increasing 'relevance' to short-term problems, to shoddy research in which, through ignorance, an old idea is re-worked to the same old conclusions, or to plodding research in which exciting results are neither expected nor wanted by the (governmental) sponsor. This is not to say that is inevitable, since much government-funded research is excellent, but this is the long-term *trend.*[10]

Short-term fears thus take precedence over long-term certainties. Genetic research is thus restricted – or at least over-regulated – because vague apprehensions that it may cause new diseases are more important than long-term certainties that it will lead to cures for all forms of cancer.[11] And in space travel, where governments do not need to show a profit, they have a strong institutional incentive to keep costs and budgets as high as possible. How terrible it would be if people launched their own spaceships, as they fly their own aircraft, and did not need government space agencies any more! How many thousands of comfortable state jobs would disappear.

But something very fundamental in the state of society is going to change within the next half-century or so. I would not have written this book if I did not believe this to be the case. Interstellar flights are going to happen because governments, in their present overweening form, are going to vanish from the Earth.

John Quincey Adams complained in vain to Congress that America did not have a single nationally funded astronomical observatory, despite the fact that Europe had more than 130. Congress persistently refused to pay for one. When it ordered the construction of a Coast Survey in 1832, it added vindictively: 'Nothing in this act . . . shall be construed to authorise the construction or the maintenance of a permanent astronomical observatory.' Today, attempts to fund projects to guard Earth from asteroid impacts that could kill millions of people are similarly constrained by the 'giggle factor' among politicians.

Chapter 3

The Twilight of the State

I believe – and hope – that politics and economics will cease to be as important in the future as they have been in the past; the time will come when most of our present controversies on these matters will seem as trivial, or as meaningless, as the theological debates in which the keenest minds of the Middle Ages dissipated their energies.

Arthur C. Clarke, *Profiles of the Future*

Like the Vandals who conquered decadent Rome, the currency traders sweep away economic empires that have lost their power to resist.

Gregory L. Millman, *The Vandals' Crown*

Paying income tax will become a voluntary activity. People will purchase the governments they want, rather than pay for governments they neither want nor need.

John Perry Barlow,
of the Electronic Frontier Foundation

Until approximately five hundred years ago, the rule of the Church was supreme in almost all aspects of society. Then came a challenge to its authority against which it could not defend itself. This was the invention of the printing press, which tore away the fabric of the priesthood and replaced it with the nation state.

Another such revolution is coming, spurred by another technical invention; this time the Internet, which sweeps away national frontiers and is likely to diminish the power of the nation state beyond recognition.

There are fascinating historical parallels between both the causes and the consequences of these two great revolutions.

This time *governments* are in the position that the Church occupied five centuries ago. It is now they, like the spendthrift popes before them, who retard the growth of wealth by maintaining vast bureaucracies that arguably mismanage almost everything they touch and do little but try to solve problems that are created by their own existence.

We are likely to see history repeat itself. Just as the Church lost its vast powers in the sixteenth century, so in the twenty-first will governments shrink into mere agencies whose role will no longer be to command and to have grand ideas which all too often turn out to be extravagant fantasies, but to offer their services for hire. And the *cause* of each social change was – and will be – technological. In each case, a new invention comes into existence that makes the old way of life redundant.

The most deadly blow to the rule of the Church came in about 1450, when Johannes Gutenberg invented printing by movable type.

Some ecclesiastics had the foresight to denounce Gutenberg's invention, correctly fearing that it would become an instrument for spreading subversive ideas.* But denunciations of printing proved just as futile as denunciations by modern politicians of the Internet. When Erasmus printed his anti-clerical *Colloquia* in 1520, the Holy Roman Emperor decreed that any teacher found using it in a classroom was to be executed on the spot.[1] But there is no record of any such punishment being carried out. Martin Luther, with equal impunity, produced a printed treatise calling the Vatican 'a gigantic bloodthirsty worm'. The Pope, he said, was a 'bandit chief whose gang bears the name of the Church. Rome is a sea of impurity, a mire of filth, a bottomless sea of iniquity. Should we not flock from all quarters to encompass the

* It is amusing to record that in 1465 the Archbishop of Mainz, who did not share these fears, honoured Gutenberg with a pension, tax exemptions, and an annual measure of grain, wine, and clothing. It is doubtful if this official would have been so generous had he known that the ultimate effects of Gutenberg's invention would include shrinking his powerful archbishopric to the status of a tourist attraction.

destruction of this common curse of humanity?'[2]

There had been a time, two centuries earlier, when the Church authorities could summon an army and massacre an entire people on the mere suspicion of heresy.* But long gone were such days, and Luther knew it. He could not be arrested, since he had powerful protectors. The only remaining course was to destroy the printing presses he was using. But in this, the outraged and fearful Church found itself helpless. It had been easy, in earlier times, to seize and burn offensive hand-written documents which, having been created very slowly, in monasteries, existed only in tiny numbers. But the new printing presses flourished everywhere, and no sooner was one closed down than another opened up.[3] Meanwhile Europe's largely illiterate population, eager to devour the printed word, flocked to schools. As William Manchester relates in his book *A World Lit Only by Fire*, a chronicle of the rise and fall of medieval theocracy:

> Outside monastery walls, the reading public was surging, though not by [official] design . . . As the presses disgorged new printed matter, the yearning for literacy spread like a fever. Millions of Europeans led their children to classrooms and remained to learn themselves. Typically, a class would be leavened with women anxious to learn about literature and philosophy, and middle class adolescents contemplating a career in trade.[4]

It was impossible that, in conditions such as these, the monolithic Church could maintain the universal sway with which it had ruled for so many centuries. And so it proved. The most noticeable consequence was the partial disappearance of the Latin tongue, in which people of rank had always published and corresponded. But the new schoolteachers knew little Latin, and taught only in the national languages of the new printed works. Thus the ancient European system in which everyone of consequence spoke the same language with all its peoples consequently compelled to share the same rituals and beliefs was finally destroyed by the new technology. As one historian observes, 'Printing and publishing

---

* As they did to the Albigensians, a people living in southern France, in the thirteenth century.

made possible the new national consciousness and promoted the rise of modern nation states.'[5]

And so to the twenty-first century. Once more the established order is threatened by new technology, this time in the form of the Internet. The most important parallel between the Internet and Gutenberg's printing presses is that both are immune from repressive government action. Just as it was futile for Church authorities to shut down one printing press when another would immediately appear somewhere else, so documents and pictures can circulate on the Internet without interference, and if necessary can be rendered unintelligible by encryption. Until recently, government agents tapping telephone lines could perfectly easily understand what they heard, but imagine them trying to decipher a piece of electronic mail that read like this:

> 90597725D5FBF6DE0F7110EB16B82BAF1C836BB68F9
> 3B373EBD4AF17982F14BAC53FF5FE9AE7E3E66EDC4

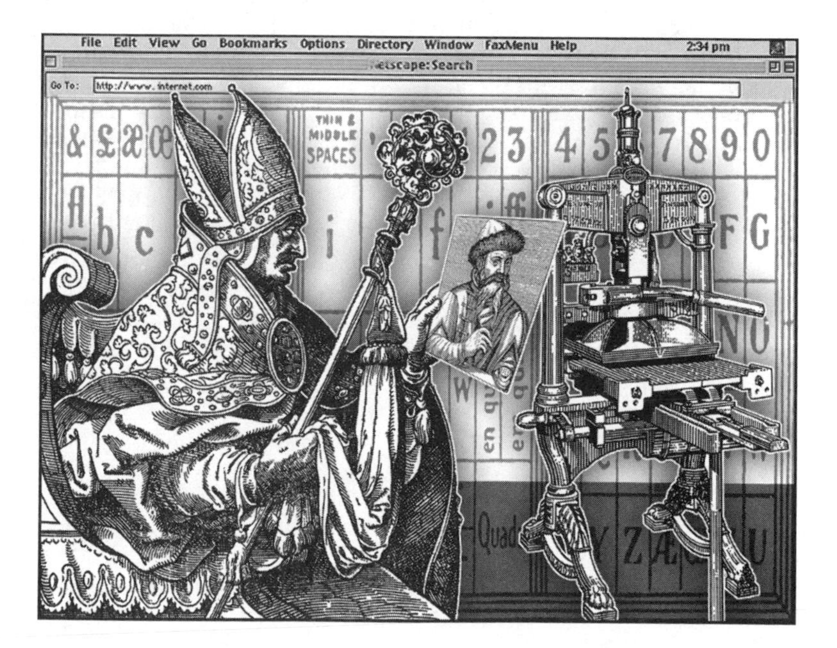

C9C75DFC8A4258855CD64F74D3DB458B6803ADF65
3665077951113C6EC652BBB4479A114E61A5D6C6864
CF0299B6FA64C6E04B0BCE1F

I encrypted the above message, which in fact means: 'Goosey, goosey gander, whither do you wander?' using the famous RSA cipher, invented in 1977 and named after its three inventors Rivest, Shamir, and Adelman. Its security rests on the sheer intractability of finding two secret prime numbers hidden inside a much larger number. The RSA is what is known as a 'public key cipher', in which someone wishing to receive secret communications can openly advertise the public part of his cipher's key in a newspaper or on the Internet, while he keeps its secret parts to himself. Without going into mathematical details here, it can nevertheless be seen that the cipher enables two people who have never met or previously corresponded to talk to each other over the Internet with no eavesdropper having the slightest ability to understand what they are saying.[6]

It might be argued that an eavesdropper has one advantage over these 'two people'. Let us call them Peter and Mary. He can identify them from their e-mail addresses. He knows the dates, frequency and length of their messages. Although ignorant of their content, he can make informed guesses about it. And if he has police authority, he can arrest them and demand information. But what if Peter and Mary decide to become anonymous? If they change their addresses, using electronic 'addresses of accommodation' that are known only to each other? I would call this 'anonymous encrypted e-mail', or AEE. (It is easy to do this on the Internet. One can have any number of secret addresses.) Now the would-be eavesdropper is truly helpless. He cannot tell who is sending the message. He cannot tell who is reading it.[7] And if a large number of citizens adopt AEE, paranoid about being spied upon by the state, the eyes of the state are blinded.

Until now, politicians have vainly wanted to 'police' the Internet on the grounds that it is sometimes used to purvey pornography. But pornography will be seen in retrospect as being the least of its threats to established order. A few among the world's tax administrations understand the threat of much more dangerous information being transmitted, that of commercial information and of *money*. Internet users can and will send this kind of data along telephone lines in encrypted form, so that tax inspectors intercepting it will encounter indecipherable cryptograms like the 'Goosey gander' one above. When this starts to happen on a sufficiently large scale, as it surely will, an extraordinary process will begin to take place. When tens of millions of people deliberately – and illegally – start to understate their incomes, because they know they can hide the truth from the tax authorities, then the 'tax base', the total amount of annual tax that a government can collect, will start to erode. The sheer size of government itself must eventually start to shrink because of insufficient funds to pay for it.

And the process will not stop there. With ever-falling incomes and ever-rising expenses, the nation state will ultimately collapse in fiscal crisis.[8]

Fragmentation of large states began even before the invention of computers, and since that time it has remorselessly accelerated. Fourteen empires disappeared during the twentieth century, and today there are powerful separatist movements in countries where the comparatively wealthy people of one region resent having to pay taxes to support what they see as the 'work-shy' of the poorer regions. An obvious example is northern Italy, where there is agitation to create the independent state of Padonia, which could become at a stroke as wealthy as neighbouring Switzerland; and in Britain, there has even been a suggestion that the wealthy Home Counties of south-eastern England could profitably declare independence. If they did this, the 18 million people in these counties would become the world's eleventh most economically powerful nation, paying no more than token taxation.[9]

Everywhere the tendency is away from large, uneconomic structures to smaller ones, and everywhere this tendency will be driven by the erosion of tax bases, caused in turn by encrypted trading on the Internet.*

Those hoping that this is a fantasy that will never happen will be alarmed to hear that the latter part of the prediction has

---

* Sometimes there will be violent reactions to this movement on the part of the states that find themselves being broken up, like the ferocious wars by Russia against its former province of Chechnya. But the sheer exhaustion caused by such civil strife, together with the expense and loss of life that it entails, may in many cases force large states to give up their territories without a struggle. Above all, there will be an *erosion of the will* of nation states to defend their right to rule over their former territories.

already started to come true. Several banks are selling computer software that enables people to trade in electronic money – money like the cash on a cash card that exists essentially in the form of IOUs.[10]

When digital money flows not merely in a cash-card economy but at lightning speed between countries on the Internet, tax evasion – that is to say, the *illegal* refusal to pay tax, as opposed to lawful tax 'avoidance' – will become as easy as pushing a button.[11]

Take the example of computer software manufacturers, who, until now, and probably for a few more years to come, have made their profits by selling their programs in the form of physical packages. These consist of cardboard boxes containing booklets of instructions and CD-ROM disks. Tax inspectors can accurately estimate the earnings of these companies by comparing the number of blank disks they have bought with those they have sold. But if, instead of selling their programs in retail shops, they invite customers to *download them via the Internet* and pay for them with credit cards by electronic mail, there is no way for the taxmen to keep a record of sales. Because no physical disk has been sold there is no record of the transaction.[12]

And what applies to software applies to *any* goods that consist of information and can therefore be encrypted and sent electronically: books, magazines, films, music, airline tickets, theatre tickets, share certificates, and legal advice. Manufacturers will understate their profits and pay less tax. They will do so with much more impunity in the cyber-economy of electronic money than with today's black-market cash economy. As one Chinese lawyer remarked of the latter: 'I made 500,000 yuan [$62,500] last year, but only told those fools at the tax office that I got 200,000. How are they going to know any better? I represent a client, someone pays in cash, and it's over. It's just less money for bureaucrats to spend on banquets, that's all.'[13]

Governments, except for a few far-seeing but helpless experts, seem unaware of what is going to hit them.* Politicians and

* Even among these experts, there is great division about what, if anything, can be done about it. The accounting firm Coopers and Lybrand advocate that governments should simply admit defeat and abolish sales tax. Others advocate either taxes on use of the Internet, or that owners of web sites should be

bureaucrats, not seeing the vast chasm that is starting to open beneath their feet, go about their business as usual, taxing and spending. In 1995, after a breakfast speech to several hundred business leaders in San Francisco, Alan Blinder, vice-chairman of the Federal Reserve, an important official of the US Government, was asked whether his organisation was studying the regulatory issues surrounding digital cash. He answered: 'Digital what?' Then, after looking puzzled for a moment, he added: 'It's literally at the thinking stage.'[14]

This answer today seems even more bizarre than when it was uttered. If the management of currency is going to be transferred into private hands while the reaction of politicians is only 'at the thinking stage', then perhaps politicians are not going to last very long.

Nor will there be much regret at their passing. There seems to be little gratitude for what Harold Wilson used to call the 'social wage', the supposed benefits that people receive in exchange for their taxes, and there is a growing number who will be happy to dispense with as many of them as they can.* As James Hale Davidson and William Rees-Mogg say in their 1997 book *The Sovereign Individual*:

> A widespread revulsion against politics and politicians is sweeping the world. You see it in news and speculation on the hidden details of Whitewater. You see it in numerous other scandals touching President Bill Clinton. You see it in reports of embezzlement by leading congressmen from the House Post Office. You see it in scandals leading to resignations in John Major's circle, and similar scandals in France, reaching two recent Prime Ministers, Eduard Balladur and

---

arbitrarily taxed irrespective of what the web sites are used for. It is hard to see how either of these measures could do anything but cause irritation rather than halt the decline in tax revenues.

* This is particularly true in the United States. According to *Time* magazine (7 July 1997), 'rarely does a town council think that Washington will provide any protection from or solution for its problems. Even if they could get federal or state aid, Americans are wary of the baggage that would inevitably come with it. If you can't help us, goes the message to government, at least stay out of the way. And it's not just militiamen wearing fatigues who are disgusted with it; alienation has joined the mainstream, fuelling tax revolts and home schooling and the growth of private-security forces.'

> Alain Juppe. Even larger scandals have been revealed in Italy, where seven-time Prime Minister Giulio Andreotti was brought to the dock on charges that included links to the Mafia and ordering the murder of Mino Pecorelli, an investigative journalist. Still other scandals have tarnished the reputation of Spanish Prime Minister Filipe Gonzales.
>
> Corruption allegations cost four Japanese Prime Ministers their jobs in the first five years of the 1990s. Canada's Justice Department alleged in a letter to Swiss authorities that former Prime Minister Brian Mulroney had received kickbacks on a C$1.8 billion sales of Airbus planes to Air Canada. Willy Claes, the secretary-general of NATO, was forced to resign under a cloud of corruption allegations. Even in Sweden, Mona Sahlin, a deputy prime minister and presumptive prime minister, was forced to resign in the face of allegations that she used government credit cards to purchase diapers and other household goods. Almost everywhere you turn in countries with mature welfare states once thought of as well governed, people hate their political leaders.[15]

It is not only corruption that people resent. There is a deepening detestation of the caste that has given us prohibition, inflation, peace-time conscription, wealth tax, double-digit sales tax, and other measures often imposed with the best of intentions but productive of the worst of consequences.* Senior politicians can do a great deal of damage even if they are not corrupt. The journalist Andrew Alexander believes that they have a tendency to go mad while in office. The very remoteness of their surroundings, he fears, must cut them off from the real world and give them vastly inflated ideas about their own importance. 'The grandeur of their position, the guards of honour, the semi-monarchical court which forms around them and plies them with flattery – soon prime ministers and presidents believe they can walk on water.'[16]

A good example of the messianic attitudes they tend to develop was given by the prominent European politician Philippe Séguin

---

* Take one of these alone: conscription. It is doubtful if a single one of the aggressive wars of the twentieth century could have broken out if dictators had not possessed conscripted armies.

when he railed against what he called 'robotisation'. The objects of his wrath were coin-operated machines at state-owned railway stations that make it easier and quicker for people to buy tickets. 'It's absurd!' he cried. 'For each machine you create three unemployed. We must stop such folly!'[17] It did not seem to interest him that if these machines were scrapped it would take longer to buy tickets and create a new layer of personnel that could go on strike and stop the trains; and that the cost of the new staff would be a fresh burden on taxpayers. Hatred and incomprehension of new technology is an increasing characteristic of modern politicians. As the computer scientist Moira Gunn put it, 'they're so blatantly uninformed, both specifically and conceptually, about technology and its impact'.[18] In the senility of their rule, politicians behave with the same reckless disregard for their own interests as the popes and their officials in the time of Luther, whose conduct included rapacity, corruption, and sexual orgies in Vatican palaces. Their self-importance today recalls the analogy in Thomas Mann's novel *Buddenbrooks* that human institutions, like stars, often shine at their greatest brilliance when their inner decay is most advanced. At the time of the French parliamentary elections of 1997, the journalist John Laughland published an article, 'No Wonder the French Loathe their Politicians'.[19]

France, from his account, must be among the worst-governed countries in the developed world. French businesses are taxed to a degree that inhibits them from competing internationally. It costs an employer 160 francs ($27) to pay someone 100 francs, and the employee must pay social charges and tax that brings his net income down to about 55 francs. A standard pay slip has no fewer than seven lines of deductions to various retirement plans, all calculated at different rates and on different bases. Companies need a court's permission to dismiss workers, which means they are reluctant to hire them in the first place. (A French judge in 1997 ordered the Renault car company to reopen a plant it had closed because it was making losses.) One in eight workers is unemployed, a rate which is twice as high among the young. French politicians, like the politicians of most countries, habitually promise to reduce taxes when seeking office, but almost always increase them when they have won it. In the 17 years following 1980, French industrial output grew by only 5 per cent, compared with 16 per cent in Germany, 21 per cent in Britain, and 57 per cent in the United States. Such a social system,

Laughland predicts, can only have one fate, 'eventual but certain collapse'.[20] A new French law has come into effect to restrict the number of hours that employees can work. Aimed at reducing unemployment by 'work-sharing', it will surely *increase* joblessness by raising industrial costs.

And from France in mid-1998 came the astonishing news that company *executives* were being arrested in the middle of business meetings for the 'crime' of working too many hours, for which they face fines or imprisonment. 'What are your names and why are you working late?' demanded members of a raiding government team.[21] It cannot be long, if such tyrannical policies persist, before companies close their doors and pull out of France, leaving an economic desert behind them.

France may be an extreme example of misgovernment in the developed world, but other governments – probably all of them in their present form – will follow its 'eventual collapse'. The problem is that state-run systems are virtually incapable of efficiency, and in the long run the people who suffer from their misrule will cease to tolerate them. Nobody is held personally responsible for what they do, and people are rarely dismissed when they get things wrong; and as a result everything is done badly.* All energy is taken up in evading responsibility.

We have seen this in the way that NASA mismanaged the design and construction of the international space station. It took more than 15 years, four major redesigns, and the expenditure of billions of dollars before any metal was even cut, let alone placed in orbit. And when the space shuttle *Challenger* blew up on its way to orbit in 1986 as a result of management incompetence, killing seven astronauts, not a single NASA official was publicly criticised, let alone dismissed or prosecuted.

In Britain, one of the most ludicrous scandals in modern times (although it involved no loss of life) was the mismanaged rehousing of the old British Museum Library in Bloomsbury, in a new building a kilometre away in St Pancras. This is an incredible story, but all too typical of what can happen in a modern state

---

* Typical was the discovery in 1997 that Britain's Ministry of Defence spends £73.50 to process the order for a 98p brass padlock, £74 to process the purchase of a £1.65 gasket, £73 for a £2.36 mobile phone charger, and £70 for a £7 car licence plate. It was reported that savings of up to £4 million could be made in the unlikely event that bureaucratic purchase procedures could be streamlined.

bureaucracy. It might seem a comparatively straightforward matter to rehouse a library, even one with more than six million books and documents, but nothing is straightforward for the British civil service. After 12 years of misapplied labour, nothing of the new library existed. Instead, there were complaints of an 'overdue and over-budget shambles' that involved 3,000 kilometres of faulty electric cabling, an absence of fire sprinklers, no storage facilities for its huge national newspaper collection, fragile shelves that were liable to collapse when books were placed on them, and a computerised catalogue filled with errors which constantly crashed. And unbelievably, to encourage unwilling customers to use the new computer system, the more accurate printed catalogues which it replaced were sent to another library, 300 kilometres away in Yorkshire.[22] It is interesting that despite many complex projects of improvement, no such scandal has ever been attached to the privately owned London Library or the Bodleian Library in Oxford, institutions that together are almost as large,

For privately managed enterprises usually flourish. When, in the eighties and nineties, Britain privatised its steel, airline, railway, gas, water, and telephone industries, the result was an explosion of profitable efficiency and vastly improved services.* Some privatised rail networks admittedly remained inefficient, but in others, trains were faster and less prone to strikes; someone wanting a telephone could get it installed in days rather than months; and you could actually drink the water without fear of feeling ill.

It has all happened before. Commerce, industry and social life in most countries at the end of the fifteenth century were suppressed and complicated by Church officials by means of myriad regulations which, as Davidson and Rees-Mogg drily remark, 'revealed that facilitating productivity was far from the minds of the regulators'.[23]

For a whole year it was forbidden to work on whatever *weekday* the most recent 28 December happened to fall – the supposed date of the Massacre of the Innocents by King Herod. Thus, if it was a Tuesday, no work could legally be done *on any Tuesday of the year*. Church officials were constantly discovering

* A performance that was marred only by the spectacle of the executives of some of the new companies greedily voting themselves inordinately large salaries.

new saints' days, during which both work and sex were forbidden. In an age when the popes were making huge profits by taxing prostitutes, sex between married couples was forbidden on Sundays, Wednesdays, and Fridays, for 40 days before Christmas and Easter, and for three days before receiving communion. In short, people were forbidden to procreate for a minimum of 55 per cent of the days of the year.[24] It seems a wonder that babies were ever born.

In short, the Church became a nuisance and a curb on progress. It was no longer seen as the benevolent authority it arguably once had been, but as a busybody. By various means in many countries it was gradually displaced as the ruling institution, to yield place to national governments. A similar change seems certain to take place in the twenty-first century; but this time it will be national governments who will be forced to yield place – in considerable part, at least – to private entrepreneurs and to powerful companies who, trading electronically regardless of frontiers, will pay the minimum of tax, if indeed they pay any tax at all, and who will in consequence be immeasurably wealthier than any global corporation today.*

Even today, the power of governments is visibly weakening under the onslaught of technology-driven currency trading. In 1997, the national currencies of several Asian countries faltered because traders felt they were overvalued as a result of reckless spending. In vain Malaysian Prime Minister Mahathir Mohamad denounced traders in terms more applicable to alien marauders in horror films. They were, he said, 'ferocious beasts, engaged in unnecessary, unproductive and totally immoral activity'. But his protests made no difference. The message was clear. His and other governments could no longer overspend without risking a collapse in currency values leading to more expensive imports, and sometimes – as happened in Indonesia – the forcible overthrow of a government after protests against the soaring prices of such items

* It will be hard to tax such companies on their profits, or their executives on their salaries, because in any traditionally legal sense, the company will not exist. It need have no corporate headquarters. It will exist only in cyberspace. It is true that its physical products can be taxed when they cross national frontiers, but the company itself can become immune from taxation. And all forms of import and export restrictions and 'economic sanctions' will also disappear, since it will no longer be possible to punish companies who break them.

as rice, sugar, and cooking oil, and against corruption by members of the ruling family.[25] As one trading expert put it:

> The currency market has delivered a loud-and-clear message to the countries of Asia. There is a higher power, the capital markets. Country after country is finding out that it does not have the autonomy and the absolute power to do as it pleases economically.'[26]

People, not surprisingly, resent suddenly being forced to pay more for their basic foods just because their rulers have embarked on some wild scheme that they cannot afford. Governments, as a result, are increasingly being seen, like the Church long before them, as a busybody and a nuisance. Their activities restrict economic growth and make it more difficult for the world's poorest people to improve their lives. By 'governments', I do not mean only the dictatorships and semi-authoritarian states, some of whom govern with the barbarity of the worst Roman emperors, but the so-called democracies as well.

It seems inconceivable that, as the powers of individuals and companies grow stronger and more independent of government through the use of the Internet and encryption, they will tolerate such oppression indefinitely. People will not pay taxes to governments which, as they will see it, exist mainly to plunder them. Instead of today's gigantic bureaucracies, they will insist upon states that are governed with the most rigorous economy, that maintain armed forces and police and such slimmed-down rudiments of the modern welfare state that cannot be managed by any private organisation, and virtually nothing else.*

The proof of this thesis lies in the fact that it is *already happening*. Government is rapidly weakening while the power of private entrepreneurs is growing stronger – a process that began many decades ago but has been vastly accelerated by the tools of electronics. It was calculated in 1997 that of the hundred largest

* One such society currently exists in Switzerland, where in exchange for an annual flat rate of 50,000 Swiss francs ($45,000), an immigrant can live as a resident without having to pay any other taxes. The state thus becomes a service provider rather than a government.

economic entities in the world, 51 were not *countries*, but *corporations*.[27] Already, when the leader of a developing nation wants an international hearing to improve the lot of his country, he typically chooses to address corporate investors at the World Economic Forum rather than delegates at the United Nations General Assembly.[28]

Consider the most dramatic humiliation of nation states and their bureaucracies in recent years: the downfall of the European Monetary System in 1992 at the hands of currency speculators led by George Soros and others. This year was the culmination of the hopes for a United Europe that bureaucrats had been describing hopefully in their position papers for decades.[29] Then Soros and thousands of other currency traders wrecked the scheme. Governments were faced with the choice of raising interest rates to intolerable levels or else abandoning the scheme. Not surprisingly, they abandoned it. They were furious with the speculators, who in turn resented their impertinence. As one of them said:

> The thing you have to remember about bureaucrats and governments is that what they'd really like is for everyone to go away and just let them do whatever the hell they like and be answerable to no one and sit on a big pedestal and announce to the little people below what's happening in the world. When the people in the world sit up and say: 'that's nonsense, you guys are wrecking the economy', they get upset. Then they start blaming speculators and traders and all these things. People are questioning their actions, and they don't like it. They'd rather pursue their nonsensical goals unfettered.[30]

As investors trade in $1.5 trillion worth of currency *every day*, episodes like this will recur, with still greater humiliations for nation states.[31] Indeed, there are many who see a dismal future for Europe's new currency adventure, the euro, as it is fully introduced over the turn of the century.

In these circumstances mankind will grow unimaginably wealthier, the rate of growth far exceeding the annual 2.8 per cent it has enjoyed for the last three decades.[32] This will be the nucleus of a new global society based on capitalism, free trade and voluntary taxation, in which dreams of voyages to the stars will one day be born.

Some politicians, sensing this trend, have threatened to 'police the Internet'. But this cannot be done. One might as well try to police the wind and rain. Built during the Cold War to survive a nuclear attack in which NATO computer communications might be put out of action, it is a network without a centre. This makes it a medium fundamentally different from any that has ever existed before. If a newspaper publishes something a government considers illegal, it can arrest the editor. But the Internet has no editor. No one owns it and no one controls it. Politicians, in their ignorance, cannot understand this. They are convinced that 'someone out there' is secretly in charge. They constantly accuse the service providers, the companies that sell the software that gives people access to it, of facilitating the spread of pornography. But the work of the service providers is merely technical. They have no more power to censor Internet information than an airline has power to censor the business ethics of its passengers. 'There is no way,' said William Giles, a spokesman for the service provider Compuserve, 'that we could possibly monitor all the thousands of discussions on our system. We have 4.2 million members. Where would we find the manpower?'[33]

Authoritarian states like China are at present building electronic 'firewalls' in the hope of screening out 'undesirable' information. This may work at present when only one in 10,000 Chinese has Internet access, making it comparatively easy to keep a watch on all of them. Indeed, one of the Chinese government officials in charge of regulating the Net, Xia Hong, boasts that he is confident of keeping control:

> The Internet has been an important technical provider, but we need to add control. The new generation of information superhighway needs a traffic control centre. It needs highway patrols.
>
> All Net users must conscientiously abide by government laws and regulations. If Net users wish to enter or leave a national boundary they will not be allowed to take state secrets out, nor will they be permitted to bring harmful information in.
>
> As we stand on the cusp of the new century, we need to – and are justified in wanting to – challenge America's dominant

> position. In the twenty-first century, the boundaries will be withdrawn. The world is no longer the spiritual colony of America.
>
> Judgement Day for the Internet is fast approaching. At most it can keep going for three to five years. The sun is setting in the West, and the glories of the past are gone for ever.[34]

Reporters from *Wired* magazine sought comments on this statement from a technician at Beijing University, a manager at China Telecom, and even another government official, and they replied with variations on the same answer: 'These people are completely out of touch with reality.'[35]

For this policy will certainly break down when economic growth has multiplied the number of Chinese Internet users a hundredfold. And it may cease working even sooner now that the more technically sophisticated population of Hong Kong, with a much higher percentage of Net users than on the mainland, has become part of the Chinese nation.*

The Chinese firewall cannot work – any more than did the original Great Wall of China, mankind's most massive work of military construction, which was contemptuously crossed in 1215 by the invading hordes of Genghis Khan. Once a user has logged on to his service provider, he need only double click on a highlighted word or title, thereby using telnet to gain access to yet another computer, leaping from network to network.[36]

I have mentioned encryption as being the one all-powerful weapon which people with computers can use against governments. Indeed, encryption could be called the 'sword and shield' of the Internet (just as the KGB used to be called the 'sword and shield' of the Soviet Union). But some far-sighted government officials believe that they have one last hope of winning back control. This lies in another technical invention likely to be made in the next few decades, the quantum computer.

---

* It may be noted in this context that, according to an estimate by the International Data Corporation, 175 million people throughout the world will be using the Internet by 2001 compared with only 28 million people in 1997. This will be an increase of 600 per cent in four years! One can only guess at the percentage of the world population that will be using it by the middle of this century, and at the social changes such use will bring.

★ ★ ★

Conventional computers, whether they are supercomputers or the PCs that most of us have on our desks, can only perform one task at a time.* This is why, if they are underpowered, they can at times seem intolerably slow. Suppose that they are trying to discover the secret keys of an RSA cryptogram, which consist of two very large prime numbers that have been multiplied together. Imagine that the two primes are *extremely* large, say about 500 digits each, creating a third number a thousand digits long when the first two have been multiplied together. Then, doing this sum *in reverse*, finding the two hidden primes, would take a network of the world's fastest supercomputers about 10 million billion billion years, a period 600 *trillion* times the past age of the universe.[37]

But in 1994, Peter Shor of Bell Laboratories showed that a theoretical quantum computer, using sub-atomic particles instead of microchips, could perform this same computation in 20 minutes.[38]

How can this be? The trick is that quantum computers (when they are built) will perform their different tasks *simultaneously* instead of *sequentially*, as in a conventional machine. Take the fairly small number, far too small to be a secure RSA key, 12,939,439,264,726,781. It will take a Pentium PC (with the appropriate software) about two seconds to discover that this number is divisible by the two primes 38,922,491 and 332,441,191. To do this, using the crudest method of computation known as the 'brute force method', it must perform about 40 million trial divisions before discovering that 38,922,491 is the smallest prime that divides into the larger number without leaving any remainder.†

---

* A single computer can *appear* to be doing two or more things at once, in a process called 'multi-tasking' or 'parallel processing'. While you are doing word-processing, for example, the computer can also keep track of the time. But this is a clever illusion. The machine is only taking advantage of very short pauses in the execution of one program during which it jumps to the other and back again.

† In the brute force method, the computer divides the number being analysed by every odd number up to its square root. There are several much faster and more sophisticated methods of doing this, but they are hard for a non-mathematician to understand. They are described in two books, David M. Bressoud's *Factorisation and Primality Testing* (1989), and Paulo Ribenboim's

A quantum computer by contrast would simultaneously explore 40 million different 'states of reality'. As one expert put it:

> Classical computers store values as binary digits (bits) that can be 0 or 1. Quantum bits (known as qubits) can represent both 0 and 1 simultaneously, so a single qubit can be involved in two calculations at once. Two qubits can carry out four operations at once, three eight, and so on. The more qubits you have, the more of a speed-up due to 'quantum parallelism' you get, leading to potentially mind-boggling performance.[39]

The quantum machine could thus reduce the task of breaking secure codes, which today demands timescales of millions of years, to just a few minutes. At a stroke, ciphers like the RSA will be rendered transparent. The Internet will have lost its sword and its shield, and governments will once more feel free to probe the secrets of its citizens.

Or so it might seem to them.

For there is a catch. What quantum electronics takes away with one hand, it gives with the other. While rendering other ciphers useless, it provides a cipher of its own that is absolutely unbreakable. In the sub-atomic world of quantum mechanics nothing can be accurately observed. The very act of observation changes it. Its laws are based on the Uncertainty Principle discovered by the German scientist Werner Heisenberg in 1927, which shows that a single particle of light can be in billions of different places simultaneously. Why? Because the very act of observing them *causes them to move in an infinite number of different ways simultaneously*. Experiments in quantum cryptography show how people wishing to communicate in secret can exploit this principle. Any attempt to eavesdrop a quantum-encrypted message – transmitted along optic fibres – instantly turns the message into unintelligible garbage.[40]

Perhaps quantum computers may never come into existence. Being based on exotic sub-atomic technology, they may prove to be fragile and hard to control. But it makes no difference. If they

---

The Little Book of Big Primes (1991), both published by Springer-Verlag. But having read them, I was no wiser than when I started.

are not made, then conventional computers will hide people's secrets from governments. And if they are, then the Uncertainty Principle will do the same. In either case, a new and uniquely prosperous global economy will come into being.

In case I have given the wrong impression, this does not mean that *all* government is going to vanish, only that its scale will be radically slimmed down. I called this chapter 'The Twilight of the State', but twilight is not the same as night. John Perry Barlow's prediction that paying income tax will become a voluntary activity does not mean that nobody will want to pay it. Only a masochist (or a criminal), or an extreme pacifist, would want to live in a city without police, or a country without armed forces. There can be few who would want to live in a society in which it was unsafe to walk in the streets unless one was armed to the teeth.*

One can envisage a two-tier society. There will be, on the one hand, the weak and the vulnerable, and on the other, the ambitious, the adventurous, the rich, and the technically capable. Both, in Barlow's words, will 'purchase the government they want' which will give them the lifestyle they want. In short, what both groups see as the essentials of the state will remain, with the useless and stifling dead hand of state bureaucracy removed.

Freed from the dead hand, the emerging free global economy, consisting of wealthy individuals rather than wealthy states, will be one that will eventually take people to the stars in the pursuit of private profit.

---

* The columnist Thomas L. Friedman interviewed some Silicon Valley executives, and found more extreme views (*International Herald Tribune*, 20 April 1998). Some of them foresaw a new world order based on electronics and information and without any semblance of government. They seemed unaware of the role of armed forces in preserving the stability that made their lives possible. He added: 'They say things like: "We are not an American company. We are IBM US, IBM Canada, IBM Australia, IBM China." Oh yeah? Well, the next time you get in trouble in China, call Li Peng for help.'

Chapter 4

The Migratory Imperative

> The leading characteristic of our species is a taste for adventure. Something was there in the mind to cry 'Forward!' though no man knew exactly whither.
>
> Robert Marett, *Man in the Making*

> For my part, I travel not to go anywhere, but to go. I travel for travel's sake. The great affair is to move.
>
> Robert Louis Stevenson, *Travels with a Donkey*

Why will people *wish* to go to the stars? The simplest possible answer is that mankind, like most animals, is a roaming species. Man has a craving to migrate. He goes *everywhere*, unpredictably, to every possible destination, provided only that it is habitable or can be made so.

This desire to wander is an instinct that must have been born among the higher primates long before man's appearance. The land masses and the ocean islands have been colonised and re-colonised, a process that started many thousands of millennia ago, both during the ice ages and in the periods between them. No natural barrier, whether of ice, tundra, sea, desert, jungle or mountains, has been able for long to halt these roamings, and it seems most doubtful that the barrier of interstellar space – vast though it is – will be able to halt them either. It may be no coincidence that in the ancient Sanskrit language, there was a common word for 'jungle' and 'desert', which merely meant barrier or obstacle.[1]

The anthropologist Ragnar Numelin sees this instinct as one of the most powerful that exists:

> There is a force which like a strong undercurrent runs through the history of mankind, or one might say through the history of all life. It is one of those deep, powerful forces which seldom stand out clearly but which are all the more striking in their effect.
>
> Movement is the name of this force, and wandering is its biological manifestation. The tendency towards movement, the source of movement, is to be found in the very organism itself. Every living organism is in perpetual motion, and the same is true of those forces which join together the human social organism. The life of man, as well as that of plants and animals, is made up of movements and counter-movements.[2]

The story of future starships is therefore only the next stage in an odyssey that has continued for epochs unnumbered. It did not begin 30 years ago with the first Moon landings, nor a century ago with the first powered flight, nor in the seventeenth century with the development of experimental science, nor even with the mathematics of the ancient Greeks. It began more than *five million* years ago, when some of our earliest ancestors who were recognisably human set out from the regions in east Africa where they had evolved to explore and settle in the rest of that continent.

The archaeologist Clive Gamble, in his book *Timewalkers*, has described the four main periods of prehistoric migrations.[3] Countless others have taken place, but these were the principal ones:

1. Between five million and a million years ago, the hominids spread themselves through eastern Africa from what is now Egypt down to the Cape of Good Hope.
2. Between a million and 200,000 years ago, people extended their wanderings to west Africa, China and the East Indies.
3. Between 200,000 and 40,000 years ago, all of Europe, Asia and Australia was colonised.
4. Between 40,000 years ago and the present day, mankind spread first across the Americas and the Pacific, and then the

> remainder of the Earth, including, in the twentieth century, Antarctica.*

And in the twentieth century, 12 men walked on the Moon, and robotic spacecraft visited every one of the Sun's planets except for Pluto. There seems to be no limit to the force of Numelin's 'strong undercurrent'.

But there is a deeper mystery. *Why* do people migrate? Scholars have put forward a host of reasons. To Jack London, author of *The Call of the Wild* and other stories about the wilderness, it was fear driven by animal passions:

> Dominated by fear, and by their very fear accelerating their development, these early ancestors of ours, suffering hunger pangs very like the ones we experience today, drifted on, hunting and being hunted, eating and being eaten, wandering through thousand-year-long odysseys of screaming primordial savagery.[4]

Fear, of course, is not the *only* driver of migration. There are also the motives of commercial gain – that drove Columbus and other European explorers of his day – of greed, desire for conquest, flight from persecution, reaction to climatic change, curiosity, and simply an appetite for adventure. As Rudyard Kipling poetically expressed this latter idea:

> Till a voice, as bad as Conscience, rang interminable changes
> On one everlasting Whisper day and night repeated – so:
> 'Something hidden. Go and find it. Go and look beyond the
> Ranges—
> Something lost behind the Ranges. Lost and waiting for you.
> Go!'[5]

---

* Apart from the many scientists who live there semi-permanently, about 7,000 tourists visit Antarctica each year. Tourism, which is probably a symptom of the migratory urge, is the world's biggest industry. According to an article in the *International Herald Tribune* (25 January 1995) it creates a new job every 2.5 seconds, and generates investments of $3.2 billion a day. By 2010, more than 900 million people (one sixth of the human race) are expected to be travelling annually for pleasure.

Indeed, all the different motives that I have listed above may be mere excuses that conceal the desire to migrate for the sheer sake of it. People feel they ought to be travelling even if they can see no reason to do so. And the mere rumour that something is hidden 'behind the ranges' – even if it is unsupported by proof – can be enough to get them started. One anthropologist, after studying the tribes of Tierra del Fuego, calls this motive the 'permanently wandering spirit' which haunts a tribe like a warning ghost if they try to ignore it.[6] The Bedouins, who have been nomads for at least 3,000 years, today lead similar lives to the ancient wandering Israelite sect the Rechabites, who made nomadism part of their religion.* God commanded them: 'Neither shall ye build house, nor sow seed, nor plant vineyard, but all your days ye shall live in tents; that ye may live many days in the land where ye be strangers.'[7]

Some peoples decide to migrate for the most frivolous reasons, or even when doing so is against their interests. The Mpongve (or Pongoue) tribe of the Gabon in west Africa – a region of tropical rain forest – are said 'never to plant two successive years in the same place, having therefore much labour in clearing the ground every time'.[8] And there is a story of a tribe in South America who decided in the middle of the night to migrate, and set out before sunrise for a march of hundreds of kilometres. Their reason: they were being annoyed by some insects.[9]

An important motive for deciding to migrate is not only to travel but to arrive. Perpetual nomadism, as practised by the Bedouins and the Rechabites, is rare. There is often a legend, or a true history, of why a tribe ceased wandering and made one place their home. The historian William Prescott described how the Aztecs – two centuries before the Spanish Conquest – arrived at what is now Mexico City, and decided to travel no more:

> After a series of wanderings and adventures, which need not shrink from comparison with the most extravagant legends of the heroic ages of antiquity, they at length halted on the

---

* Long-distance travellers, whether they are nomads or astronauts, can exert a fascination over more sedentary people. There is a story in Israel of a wealthy American tourist who rejects offers to show him temples and churches, saying instead: 'All I wanna see is a genoooine Bedoouine Ayrab.'

> south-western borders of the principal lake. There they beheld, perched on the stem of a prickly pear, which shot out from the crevice of a rock that was washed by the waves, a royal eagle of extraordinary size and beauty, with a serpent in his talons, and his broad wings opened to the rising sun.
>
> They hailed the auspicious omen, announced by an oracle as indicating the site of their future city, and laid its foundations by sinking piles into the shallows. On these they erected their light fabrics of reeds and rushes; and sought a precarious existence from fishing and from the wild fowl which frequented the waters. This place was called Tenochtitlan, in token of its miraculous origin. The legend of its foundation is still further commemorated by the device of the eagle and the cactus, which form the arms of the modern Mexican republic.[10]

It is interesting that a significant amount of the world's literature concerns long journeys, or 'quests'. Odysseus spends ten fascinating years trying to return home. Captain Ahab combs the seven seas for the white whale. Countless gallant knights search the world for the Holy Grail. Julius Caesar's *Commentaries*, a best-seller among the contemporary Roman populace, describe a victorious general in almost continuous motion. He is constantly *arriving* at a battlefield by the most unexpected route, his enemies imagining he is hundreds of kilometres away. Two of Conrad's novels, *The Nigger of the Narcissus* and *Heart of Darkness*, are about long and perilous voyages. Alexander the Great, whose whole adult life was one extended military expedition, always carried in his travelling chest a copy of Homer's *Odyssey*, a tribute to one wanderer from another. Perhaps to travel can be more interesting than to arrive. Hundreds of pages in *The Lord of the Rings* are taken up with Frodo's journey to Mordor, but only one page with what he does when he completes it. One of the finest narrative poems of the nineteenth century, Robert Browning's *How They Brought the Good News from Ghent to Aix*, describes the perilous dash of three intrepid horsemen. The only thing the author fails to tell us is what the good news *was*. He does not seem to think it would interest us.

Sometimes there is a background of heroism to people's decision to migrate. They migrate because of discontent and despite any fears of what they might encounter at their destination. A good example of this was the oppression that drove a huge number of Puritans across the Atlantic from England during the reigns of the first two Stuart kings. Best known among these departures was the voyage of the Pilgrim Fathers in the *Mayflower* in 1620. Grim was the debate as they pondered whether to stay in Europe where they had no prospect for advancement, or face the appalling dangers of the New World. As their leader William Bradford recorded it:

> They fixed their thoughts on those vast and unpeopled countries of America, which were fruitful and fit for habitation, though devoid of all civilised inhabitants and given over to savages, who range up and down, differing little from wild beasts.
>
> This proposition raised many fears and doubts. The hopeful ones tried to encourage the rest. Others, more timid, objected, alleging much that was neither unreasonable nor improbable. They argued that so big an undertaking was open to inconceivable perils and dangers. Besides the casualties of the seas, they asserted that the length of the voyage was such that the women, and other weak persons, could never survive it. And the miseries which they would be exposed to in such a country would be too hard to endure. The change of air, diet and water would infect them with sickness and disease.
>
> Even those who surmounted these difficulties would remain in continual danger from the savages who are cruel, barbarous, and treacherous, furious in their rage, and merciless when they get the upper hand. Not content to kill, they delight in tormenting people in the most bloody manner possible; flaying some alive with the shells of fishes, cutting off the members and joints of others piecemeal, broiling them on the coals, and eating collops of their flesh while they live – with other cruelties too horrible to relate. [11]

But it is the Polynesians, whom I mentioned in Chapter 1, who must serve as the best model for future interstellar explorers and colonists. If, instead of thinking of stars and planets, we imagine tens of thousands of habitable islands, we will have a good historical parallel.* Reduce the scale of distances from light years to hundreds of kilometres, and consequently the speeds needed to reach one region from another, and the model may be almost exact. Like space, the Pacific contained no barriers. It was a 'highway that favoured intercourse and migration'.[12] As an explorer said of the South Sea Islands after a voyage in the 1830s: 'Exactly twice the extent of the ancient Roman Empire in its greatest glory, the same primitive language is spoken, the same singular customs prevail, the same semi-barbarous nation inhabits the multitude of the isles.'[13]

The colonisation of the Pacific by the Polynesians was one of the most tremendous enterprises in history. It was carried on for century after century. Why it began, what motive prompted the migrations of these ancient peoples from Asia – or, as Thor Heyerdahl has argued, from South America – no one knows.

They had no metals and no system of writing. Their tools were made from shells and from plants and vegetables, their ropes from hibiscus trees, and their sails of coconut fibre. They had no navigational instruments, only generations-old instincts and techniques. Like future interstellar travellers (see Chapter 12 'The Star on the Starboard Beam'), the stars were their guides. And they could tell if they were approaching land, even though it was still beyond the horizon, by the changing patterns of the waves.[14]

By 1500 BC, three centuries before the Siege of Troy, they occupied Fiji, Samoa and Tonga and all the islands east of the Solomons. By AD 1000, when Europe was deep in the Dark Ages, they had made unimaginable advances. They had filled every niche where human life could prosper in the vast oceanic triangle made by Hawaii, Easter Island and New Zealand, a region of some 18 million square kilometres.[15] But these were no haphazard voyages. There was purpose behind them, as Jared Diamond explains:

> Historians used to assume that the Polynesian islands were discovered and settled by chance, when canoes full of fishermen happened to get blown off course. It is now clear, however, that both the discoveries and the settlements were

---

* In using the term 'Polynesia' to describe its islands and peoples, I include those also of Micronesia and Melanesia, as do most historical writers on the subject.

meticulously planned: since much of Polynesia was settled against the wind and currents, it is unlikely that voyagers arrived by drifting. Furthermore, they carried with them many species of crops and livestock deemed essential to the new colonies' survival, from taro to bananas and from pigs to chickens – a transfer that was certainly deliberate.[16]

It is the 'radiative' nature of the Polynesian voyages that provides the closest parallel to interstellar travel. It was all made possible by that forerunner of a starship, the double canoe. Imagine twin hulls about nine metres in length, covered by a single deck and a lateen sail, a craft not dissimilar to a modern catamaran. These vessels averaged about 12 knots when the wind was favourable, enabling them to cover distances of 3,000 kilometres in a week, making possible the settlement of islands at the rate of more than a hundred in a year.*

Now these vessels were 'self-reproducing', in the sense that they were built wholly from local materials. Each time a colony settled on an island, they would cut down trees to make fresh canoes. Each of these canoes would set forth to find fresh islands, and on arriving on these, each party would eventually make fresh canoes, and so on.[17]

It is by a method akin to this that our descendants will learn to 'hop' from star to star, finding suitable planets as they go, always using local materials for the construction of colonies. For migration in the future in space will be but a continuation of migration on this planet in the past, a continuing exercise of what might be called the 'migratory imperative'.

We have seen how mankind has spread to fill every available niche on one planet. Driven by this imperative, he will surely do the same, when the technology is available, to at least part of our Milky Way galaxy.

The chances of competition from any alien civilisations, at least in this local region of the galaxy, seem remote. Both observation and theory support this view. Many years of telescope searches have revealed no signs of artificial radio signals. An important announcement to this effect was made by the radio astronomer Paul Horowitz in 1993. He and his team had spent the previous eight years (an exercise that still continues) listening for radio signals that might have been sent to us by alien civilisations among the nearby stars. But the results have been wholly negative. As Horowitz explained:

> We have picked up more than a million radio signals which, at first sight, might have been artificial. But most of them proved to be computer errors or just random noise. We at last narrowed down the search to thirty-seven radio signals which cannot be explained. They are not the result of

---

* Today, South Sea islanders think nothing of crossing the ocean for five days in order to buy cigarettes.

> equipment failure, nor do they originate from aircraft or Earth satellites.
>
> But they all fail one vital test: they do not repeat. In none of the thirty-seven cases could we pick up the same signal twice. This failure violates one of our most important assumptions, that an advanced civilisation would be intentionally transmitting to us. A real signal would therefore consist of a continuous stream endlessly repeating the same message.[18]

Just as significant are the views of the theorists. There is a growing consensus that if an advanced society of aliens *did* exist they *also* would have spread to every available niche, which would include Earth. It seems reasonable to assume that they too – some such societies at least – would be possessed by a 'migratory imperative'. But there is no sign that this has happened. Thus, since no alien artefacts have ever been found on Earth, there is a strong case for believing that alien civilisations technically more advanced than our own do not exist, at least not locally.[19]

To put this another way: the Sun is a second-generation star, which means that our solar system is only half the age of the galaxy. If, therefore, an alien civilisation from one of the first generations of suns had set out to colonise the galaxy, they would have had several billion years in which to do it. They would have colonised every habitable star system, including ours. The fact that none appears to have done so is further evidence that there are no advanced alien civilisations, at least not in this galaxy.

The apparent absence of inhabited planets in no way precludes the existence of *habitable* planets, and that means planets – like the hypothetical one I described in Chapter 1 – of the same approximate size as the Earth. One planet-hunter, Andrew Lyne of Jodrell Bank, believes that 'on a very conservative estimate, we can now speculate that one in a hundred stars in our galaxy may have Earth-sized planets. If this is true, then there are a billion such worlds in the Milky Way.'[20]

Reflecting on these figures, one can only decide that few beliefs seem more ridiculous than that the only intelligent race among billions of stars should confine itself for ever to one planetary system. At the risk of sounding mystical, it would be a betrayal of our destiny, especially since for far more than 99 per cent of the future history of the universe there will be no planets to go to, because atoms themselves will have disappeared (see Chapter 20, 'Extinction and Eternity').

Yet interstellar flight will not be like flying to Mars – a feat that should be comparatively easy in the early part of the coming century. In the words of two engineers who have studied the problem, it is 'not just hard, it is very, very, very hard!'[21] Small wonder that many scientists, astronomers in particular, who know a great deal about distance but not very much about man's migratory urges, dismiss the whole idea as impossible. Their sweeping statements remind us of another, by the New Zealand scientist A.W. Bickerton, who declared in 1926:

> This foolish idea of shooting at the Moon is an example of the absurd lengths to which vicious specialisation will carry scientists working in thought-tight compartments. To escape the Earth's gravitation a projectile needs a velocity of eleven kilometres per second. The thermal energy at this speed is 15,180 calories [per gram]. Hence the proposition appears to be basically impossible.[22]

But interstellar flight will still be 'hard' with all those 'verys'. For the distances between the stars are truly vast. They are on a scale a million times greater than the distances between our local planets, and much of this book is devoted by necessity to discussions of tremendously high speeds and exotic methods of propulsion.

A simple exercise will illustrate this idea. Imagine that the Sun, in reality nearly one and a half million kilometres in diameter, was reduced to the size of a metre and placed in the middle of London. The Earth, on this scale, would be a ball about the size of a grape pip a city block away. Jupiter, the size of a grape, would be half a kilometre from the miniaturised Sun. Alpha Centauri, comprising the Sun's closest stellar neighbours, would be in Canberra.*

---

* Much more extreme miniaturisations would be needed to confine the entire universe into a truly manageable volume. To make the whole cosmos fit into a locket that could hang round someone's neck, it would be necessary to shrink the Sun down to the size of a quark, the smallest known atomic particle, with a diameter of only 10^{-15} centimetres.

Now return to the real universe. Circling the Sun along with our world are a host of interesting and resource-filled planets, asteroids, and moons. Beyond them, extending out to a distance of about a light year, lies the vast Oort Cloud (after Jan Oort, who proposed its existence in 1950) – of some 10 trillion comets. Beyond these lies apparently nothing except for wisps of gas and dust, until just over four light years, when we reach the nearest stars. These will be the future equivalent of the Pacific islands.

We have one advantage over the ancestors of the Polynesians at the time when they contemplated the seemingly boundless vastness of the Pacific. We have no horizons to prevent us from seeing where we are going. We shall not flinch from exploring, any more than they did. Within perhaps less than two centuries from now we will have the capability to build starships.

In this era of ever-advancing technology, it would be ridiculous to suppose that we would go so far in constructing ever more powerful machines and then, unaccountably, stop. The moment will come, perhaps in the lifetime of our great-grandchildren's great-grandchildren, six generations hence, when we will have exploited and colonised all the useful planets, moons and asteroids in our solar system, and must go further or else suffer whatever stagnation is brought about by suppressing the migratory imperative.

The great Russian visionary Konstantin Tsiolkovski, who wrote his first paper on astronautics in 1903 – the very year in which the Wright brothers were making the first powered flights – dreamed of everlasting migration. To him, the occupation of all Earth was only the beginning. 'The finest part of mankind,' he wrote, 'will in all likelihood never perish. They will migrate from sun to sun. And so there is no end to life, to intellect and the perfection of humanity. Its progress is everlasting.'[23]

A final reflection. Tsiolkovski may have been right in principle but wrong in detail when he spoke of interstellar migration by 'the finest part of mankind'. If he meant – as he appears to have done – that these migrants for the most part will be the rich, the adventurous, the profit-hungry and the dreamers, then I believe he will be proved wrong. For although this category of people will be the organisers of the voyages and the entrepreneurs who will benefit from them, they cannot constitute the bulk of the colonists. No great wave of migration could be carried out by people such as these. They would have no aptitude for the hard, and often dull work that must be carried on for several generations if a new civilisation is to be successfully planted. They would be more interested in contemplating new enterprises than in draining marshes. In a word, they would be too 'grand' in their thinking for such mundane labours.

Instead, I suggest, the majority of the interstellar colonists will be poorer people who feel that continued life on Earth offers them no chance to better themselves, either because they feel imprisoned by unfavourable natural environments, or else because they live under oppression. As one scientist put it:

> If you offered the chance to spend the rest of their lives on another planet that was much like Earth to people under thirty from a poor family in Africa or South America, or to a farmer in Bangladesh who lived in perpetual fear of losing his crops and his livelihood from the floods from the Bay of Bengal, they'd jump at it. I believe they'd be prepared to take any reasonable risk. True, they would be unable to pay for their passage, but they would make agreements of indenture, obliging them to work, for a certain period, on their planet of destination. They would go, not from any idealistic reasons, but because they saw no hope in their present conditions. [24]

One can see from the past that in many cases it was just such motives that drove people to set out for distant continents. Either they were discontented, or else they were persecuted, and sometimes there was a combination of the two. Above all, when these feelings were strong enough, there was a contempt for danger. I have already mentioned the Puritans and the *Mayflower*, and so I will remark on one other striking example of a migration of this kind.

It took place from France in 1685 with the revocation of the Edict of Nantes by Louis XIV, which abolished the religious freedom of the Huguenot Protestants. Nearly half a million Huguenots – perhaps a tenth of the population – gave up their livelihoods rather than change their religion. Tens of thousands of them undertook perilous journeys to North America where they settled in New York state, Virginia and South Carolina, and in Canada, where they found themselves facing frightful wars against the Iroquois Indians.[25] Everywhere the exiles were a source of strength to the countries that received them. French officers drilled the Russian armies of Peter the Great. Three Huguenot regiments were added to the English armies, and Huguenots established themselves in England as silk workers, cotton spinners, wool carders, hat makers, weavers, and makers of sailcloth, glass and paper. An act of oppression had dispersed an entire people across the globe.

We may never be able fully to answer the question, why do people migrate? Perhaps it is unanswerable, because there are so many different motives. But this seems plain: if there is somewhere to migrate to, then some will want to make the journey.

A more practical question is: to which alien sun will our descendants first venture?

Chapter 5

The Story of Rigil Kent

> At last in sight the Centaur drew,
> A mighty grey horse trotting down the glade.
> Over whose back the long grey locks were laid,
> That from his reverend head abroad did flow;
> For to the waist was man, but all below
> A mighty horse, once roan, but now well-nigh white
> With lapse of years; with oak-wreaths was he dight
> Where man joined onto horse, and on his head
> He wore a gold crown, set with rubies red.
> And in his hand he bare a mighty bow,
> No man could bend of those that battle now,
>
> William Morris, 'The Life and Death of Jason'

This personage was Chiron, the only respectable member of the clan of centaurs, mythical beasts of ancient Thessaly, who were reputed to be horses with the heads and torsos of men. The myth probably originated in a local tribe known for their expert horsemanship and wild behaviour. The story goes that they were invited to a wedding feast at which they tried to abduct the bride. The battle which this action provoked with the other guests, leading to the expulsion of the centaurs from Thessaly, was commemorated in engravings on the Parthenon and on innumerable vases.[1] (An original depiction of it can be seen in the British Museum.) Chiron, the supposed tutor of Hercules, who attempted to restrain his comrades at the wedding, was immortalised by being made the brightest star of the constellation of Centaurus, Alpha Centauri.

Alpha Centauri is in many respects the most interesting star in the sky beyond the Sun, as well as being the most important in mankind's eventual future. As one expert puts it:

> My guess is that Alpha Centauri will become somewhat equivalent to the Americas in the exploration of the Earth. At first it will be a distant outpost, but over time it will grow into a civilisation on a scale commensurate with its resources. Ultimately, it will come to dwarf the Old World. Our original solar system will always be revered as the ancient home of life and mankind, but the Alpha Centauri system will eventually become the epicentre of galactic colonisation.[2]

Let us look at the characteristics of Alpha Centauri, and the surprises it has given to astronomers through the ages, a study that will show the importance of its 'resources'. One of the four stars of the Southern Cross, its most important resource is its proximity. It is the nearest star to the Sun, only 4.3 light years from Earth, contrasting with the remotest stars in our galaxy, which lie 100,000 light years away.*

Consider how advantageous this will be for our descendants. The average distance between star systems in this region of the galaxy is between seven and eight light years.[3] But even hypothetical beings with the nearest star eight light years from their own are fortunate compared with some. Amid the Virgo cluster of galaxies that lie some 50 million light years from Earth there are 'lone stars', stars without neighbours that do not belong to any galaxy, which were long ago flung out of their parent galaxies, probably by the gravitational tides of galactic collisions.[4] If any beings live on planets circling one of these lonely stars, their night skies will be black indeed. Their nearest stellar neighbours are *millions of light years* away, and they might imagine their parent sun to be alone in the universe. Dark indeed would be their future prospects. The idea of interstellar travel would probably never occur to them.

Alpha Centauri, by contrast so encouragingly close to us, is also by far the most massive object among the stars that lie in the Sun's immediate neighbourhood, a region about six light years across. If the Sun were seen at the same distance, the light of Alpha Centauri would be four times greater, and its mass would be double that of our entire solar system of Sun and planets. This is because while it appears through the naked eye to be a single star, it is in fact a system of three stars.

For thousands of years Alpha Centauri has for some reason had far more names than any other star. It is commonly named Rigil Kent, meaning the brightest star in Centaurus (the term Kent, or Centaurus, is added so as not to confuse it with Rigel,

---

* Although part of the Southern Cross, Alpha Centauri lies outside the neighbouring constellation of that name, the Crux Australis. The constellation of Centaurus was known to the ancients, but part of it, containing three out of the four stars of the Cross, was named in 1503 by the explorer Amerigo Vespucci – after whom America is named. The Southern Cross today appears in the national flags of Australia, New Zealand and Western Samoa.

the brilliant star in Orion). Indeed, to call it 'Alpha Centauri' reflects a certain northern bias. We do not normally call Sirius, the brightest star in the whole sky, Alpha Canis Majoris, which it technically is. There is a strong case for giving bright and prominent stars familiar, non-technical names.[5]

Being visible high in the sky anywhere between North Africa and the southern oceans, Rigil Kent was wondered at by many ancient peoples. It became an object of worship on the Nile, and the ancient Egyptians made it a god which they called Serk-t, using it to orient the positions of at least nine temples. In China it was the celestial 'Gate to the South', or Nan Mun. It has also been called Bungula, meaning the hoof of the centaur, and, mysteriously, Toliman, which may have been the name of a ship's captain in the days when captains navigated by the stars.[6]

Being a southern star, Rigil Kent has always attracted much less attention than celestial objects visible from the northern hemisphere, the abode of most astronomers for thousands of years. It was never even seen by Europeans until the age of exploration began in the fifteenth century. Its remarkable characteristics were only gradually discovered thereafter. In 1689 a Jesuit astronomer named Father Richard was searching for comets in Pondicherry, India. One night, wearying of this quest, he turned his telescope on Rigil Kent. It must have been a temptation that was hard to resist, since compared with the faint comets he had been hunting for, Rigil Kent was tantalisingly bright in the south. Yet it was well above the horizon, since Pondicherry is at the 12th latitude above the equator. He found to his astonishment that it consisted of *two* stars while through his naked eye he could see only one. This was the first double star system to be discovered, and the two companion stars are now called Rigil A and Rigil B.[7] It was a prophetic discovery. As mentioned earlier, it is now known that perhaps half the stars in our galaxy, at least in regions of it near the solar system where we have been able to observe them, consist of double or triple stars. Our Sun, without a stellar companion as far as is known, is a rare exception.*

---

* There is an unconfirmed theory that the Sun *does* have a companion star, far off and too faint to be easily discovered. It has been called Nemesis, the 'death star', since its supposed close approaches to the Sun, occurring every 26 million years, might disrupt the billions of comets that surround the solar system. Many of

The next great discovery about Rigil Kent, which was also epochal to the subject of interstellar travel, was made in 1832. Its author was the lawyer-turned-astronomer Thomas Henderson, who was partly blind and a sufferer from heart disease which makes his achievement all the more remarkable.[8] Appointed in that year to direct the Royal Observatory at the Cape of Good Hope in South Africa, he immediately turned his attention to this star – or, one should say, this pair of stars. He was not interested in their stellar composition, but in their *distance*. Nobody at that time had any idea how far away any of the stars were. Even the great Johann Kepler, two centuries before, had inclined to the ancient Greek view that all the stars were the same distance from the Earth, probably a fairly small one, and that they all lay in a shell surrounding the solar system that was only a few kilometres thick.[9]

Henderson had the instruments to do what no one had ever succeeded in doing before. He observed the exact position of Rigil Kent in the sky. Then he waited precisely six months until the Earth was on the opposite side of the Sun. He again took Rigil's position, and found that since his last observation it appeared to have moved in the sky by the tiny amount of three-quarters of a second of arc.

In effect he was imitating a man who conducts a simple experiment with his eyes and thumb.[10] Hold out your hand with fist clenched at arm's length with your thumb pointing upwards. With right eye closed, take a sighting past your upturned thumb at a distant object. Then open your right eye and close your left, and again take the sighting. Your thumb appears to have taken a tiny jump from left to right. For thumb, read star; and its changing position, depending on which eye is closed, is dictated by where the Earth is in its orbit round the Sun. Henderson knew that in the six months between his two observations, the Earth had traversed a distance of 300 million kilometres. This distance formed the base line of a triangle. Trigonometry thus enabled him to work out from the distance between the two apparent positions of the star its distance from the Sun. It turned out to be 275,000

---

them would crash into the Earth, and this would explain the periodic mass extinctions of life that have occurred in pre-history. The search for Nemesis continues.

times greater than the distance between Earth and Sun. This amounted to 41 trillion kilometres, or 4.3 light years.*

It was not until 1915 that the most astonishing discovery – at least it seemed astonishing at the time – was made about the Rigil Kent system. Robert Innes, a former wine merchant turned meteorologist who practised astronomy in his spare time, discovered that it was not a system of two stars but of *three*. In contrast to its two brilliant companions, the third star was 300 times fainter than the faintest star that can be seen with the naked eye. Using a one-metre telescope at the Transvaal Observatory in Johannesburg, Innes compared a picture of the Rigil Kent system that had been taken in 1910 with one that his assistant took in 1915.

He used a method called the 'blinker', a technique at which he was a pioneer. For this, he was criticised by conservative colleagues who preferred the more conventional astronomical method of merely taking photographs of stars and then examining them visually.[11] The blinker, or blink comparator, versions of which are still being used, was a primitive kind of computer. Two photographs of the same object, taken months or years apart, are mounted on a device that compares them, detail by detail. Any difference in the two pictures, however slight, jumps out at the viewer. Innes in 1915 discovered such a jump in his second picture of Rigil Kent. From this he deduced the presence of a third star that was 0.1 light years or nearly a trillion kilometres closer to us than the other two stars. It is thus the nearest star to the Sun and has since been called Proxima Centauri.[12]

Innes had used the same method to find Proxima Centauri as Clyde Tombaugh did to find Pluto, the Sun's most distant planet, in 1930. But in this there is a most curious story. Tombaugh became a hero in scientific circles for his achievement in finding the Sun's most distant world, while Innes, who discovered the Sun's nearest stellar neighbour, remained almost entirely unknown. In none of the astronomical textbooks I have read is his name mentioned. Tombaugh gets two lengthy paragraphs in

---

* It will be obvious that the more distant a star is, the smaller its movement will be between six-month periods. Therefore this method of calculating distances to the stars, known as 'trigonometric parallax', only works if the star is less than about 160 light years away. Beyond that, more indirect methods have to be used.

the 1994 *Encyclopedia Britannica*, while the name of Innes does not even appear in the index. Even Innes's obituary, in 1933 in the astronomical journal *The Observatory*, does not mention his discovery of Proxima Centauri!

One science writer, Ken Croswell, has discovered the strange reason for this discrepancy in fame: unlike many scientists, Innes suffered from excessive modesty. *He never used his first name in scientific papers*, only his initials. He signed himself 'R.I.'. I have his paper in front of me as I write, with a signature whose brevity has condemned him to an undeserved obscurity.[13] Since most editors hate initials and insist on first names, there is always a temptation to delete the discoverer's name and record only the discovery. As Croswell explains:

> If all I have to go on is the first letter of the first name, how do I know whether the scientist is male or female? The answer, of course, is that I don't. Which means I have to write without pronouns – which is no fun, and means I can't talk about 'his' findings or 'her' observations. Things like this drive writers crazy. It may even cause the harried writer to delete any mention of the scientist. In my case, I could have said that Proxima Centauri had been found in 1915, and left it at that.[14]

And nearly all subsequent writers about the discovery of Proxima Centauri did exactly that. From now on I shall call this third star of the system Rigil C.

There is an uncanny resemblance between the Sun and the two bright stars of Rigil Kent. Rigil A and Rigil B seem to have the same spectral signature as the Sun, suggesting that they are composed of the same elements as the Sun and that they were created at the same time. It may be that all three were created by the same supernova explosion five billion years ago. They even share with the Sun the unlikely ratio of one atom of iron to every 31,620 atoms of hydrogen, a ratio that no other star is known to possess.[15] It may be that in seeking to travel to it we may not be going to an 'alien' solar system at all, but one that is in some respects part of our own.

Are there likely to be habitable planets in orbit around either or both of them? Although we shall certainly know the answer to this question long before manned spaceships set out for them, it is useful to speculate about the matter now.

There is, unfortunately, what Charles Pellegrino calls a 'self-perpetuating textbook dogma' that states that multiple star systems like Rigil Kent cannot have planets in stable orbits.[16] Nearly every class of astronomy students is taught this, and it is nonsense. It is wrongly assumed that the gravitational field of one of the stars will catastrophically disturb the orbits of the others. This may indeed be true if the stars are very close together, within, say, about twice the distance between the Earth and the Sun. In the vigorous phraseology of Sir John Herschel more than a century ago (writing at a time when it was thought that all planets harboured intelligent beings):

> Planets, unless closely nestled under the protecting wing of their immediate superior, the sweep of their other sun, in its perihelion passage [closest approach] round their own, might carry them off, or whirl them into orbits utterly incompatible with the conditions necessary for the existence of their inhabitants.[17]

But if they are sufficiently far apart, there is no reason to suppose that there will be any such disturbance. On the scale of our solar system, Rigil A and B are *very* far apart. Although the distance between them varies as they orbit one another, its *minimum* is 11 Earth–Sun distances, nearly two billion kilometres. In other words, if we orbited Rigil A at the same distance that we orbit the Sun, Rigil B would lie somewhere out between the orbits of Saturn and Uranus.

What difference would this make to our existence? None at all. Since Rigil B is slightly less massive than the Sun, at this distance it would not affect Earth's orbit, nor would it affect life on Earth. It would simply be too far away. The only difference we would notice would be the presence of an extremely bright star in our night skies. The same would apply to conditions on a planet orbiting Rigil B. Rigil A would merely be a very bright star in *its* night skies. The two stars, in other words, are sufficiently far away from each other so that habitable planets could orbit *both* of them. Indeed, this has been proved mathematically. Computer simulations have shown that a planet circling a star of approximately the same mass as the Sun will have a stable orbit *so long as another star of similar mass never approaches the planet closer than three and a half times the planet's orbital radius.*[18] And with the two main stars of the Rigil Kent system never approaching each other closer than *11* times an Earth-type planet's orbital radius, there is no reason to assume, as too many scientists have somewhat carelessly done, that this system of stars, or systems like it, must be barren of habitable worlds.

But why all this speculation? Why should we need to debate this matter at all? Why not simply point a powerful telescope in the southern hemisphere at Rigil Kent and *see* if it has planets suitable for habitation? The answer is that this cannot yet be done. We do not have the technology. The stars are too far away for the reflected light of any of their planets to be seen. A Sun-like star typically emits a billion times more light than any planets that may lie in its orbit, and this radiation blots out the light of the planet.

Nevertheless, some 20 planets have so far been found circling other stars.

The existence of all of them, save one, has been deduced from the minute movement of their parent stars. But one alone has been actually photographed, by the Hubble Space Telescope. This 'planet', if it is properly speaking a planet at all and not a brown dwarf, is at least three times the mass of Jupiter. But the greatness of the achievement in seeing it is its distance from Earth of no less than 450 light years – its light started on its journey towards us when Henry VIII was king of England.[19] This is an incredible distance at which actually to see so comparatively tiny an object.

These discoveries are a technical triumph of the late twentieth century, an extraordinary advance from earlier times, when astronomers despaired of ever being able to find such worlds or knowing whether they existed. 'In this state of ignorance,' one of them lamented in 1859, 'to assert that the stars have, or that they have not, attendant planets, would be alike rash.'[20] But now, nearly 400 years since Galileo challenged the view of the Church that the Sun goes round the Earth, we can take advantage of the fact that in a narrow sense the Church was right. Gravity works both ways. The Earth minutely attracts the Sun, so that the Sun is very slightly moved by Earth and its sister planets. The Sun is not the absolute centre of the solar system; it actually orbits a common centre of gravity just outside its own surface.[21] Thus a star with a planet in its orbit will be seen to 'wobble' very slightly (not the *apparent* movement seen and sought for by Thomas Henderson, but a real motion). By this indirect means astronomers have been able to detect the existence of most of these alien worlds.

But since our ability to observe these tiny motions is at present so limited, it has so far been possible only to confirm the presence of giant planets, like our own Jupiter, which probably, being made of gas and having gigantic surface gravity, would be totally unsuitable for habitation.

Yet there is every hope of finding Earth-sized worlds when techniques improve, and there are strong hints that, although so far invisible and undetected, they are overwhelmingly likely to exist. One of the most interesting recent discoveries was made in 1996 by the astronomer George Gatewood. He detected the presence of a Jupiter-sized planet in orbit round Lalande 21185, the fifth closest star to the Sun, which lies eight light years away in the constellation of Ursa Major.* This planet orbits at about twice as far from Lalande 21185 as the Earth does from the Sun.

It was interesting because the distance between the planet and the star is about half Jupiter's distance from the Sun. If Lalande 21185 was the same kind of star as the Sun, Gatewood's planet would certainly disrupt the orbit of a planet at the Earth–Sun distance. But this star is a very different kind of star from the Sun. It is fainter, redder and smaller, emitting only about a third of the light that the Sun does.

This discovery thus strengthens our belief in the existence of Earth-like planets. All that is needed is that such worlds should receive a similar amount of solar heat that the Earth does. And this will be true, in the case of a dim, red star like Lalande or Rigil C, *if it is much closer to its parent star than the Earth is to the Sun.* Indeed, it has been calculated that such a world would be habitable if it circled Lalande at a distance of only nine million kilometres, one-sixteenth of the Earth–Sun distance.[22] Colonists reaching Lalande might at first complain of the 'lurid red light' of their new parent sun, but Lalande being a comparatively cool star, they would receive the same natural warmth from it as we do from

---

* Its strange-sounding name means only that it is the 21,185th star in a catalogue of more than 47,000 published in 1801 by the French astronomer Joseph Lalande.

ours. And the example of Lalande shows that a habitable world could easily have been formed in the orbit of Rigil C. There is a definite pattern of how worlds form around a star – although there will no doubt be many exceptions to it. In the words of David Black, director of the Lunar and Planetary Institute in Houston:

> Planets will be found around almost all stars. The orbits of the planets will lie close to one plane. Why? Because the whole momentum formed from a protoplanetary nebula has been conserved. The heavier planets will orbit at about Jupiter's distance from the Sun, since they formed from the most abundant chemical elements in the mix, and those elements are relatively light. Most stars will have planets in the 'habitable zone'. There is thus reasonably high probability that one planet in a planetary system will resemble Earth.[23]

The kind of planet that orbits Lalande 21185 is of great importance in making planetary systems habitable. For being as massive as Jupiter – there would be no possibility of detecting it with current technology if it were very much smaller – it cleans up comets, acting as a kind of local cosmic hoover. Not being habitable itself (one cannot live on a gas giant, for it has no solid surface), it ensures that Earth-type planets are free from constant bombardment by comets and asteroids. Indeed, we owe our own comparatively safe existence to the gas giants Jupiter and Saturn. If they did not exist, we probably wouldn't be here. In the words of the astronomer George Wetherill:

> Without a Jupiter-sized world in our planetary system, collisions with large comets and other dangerous objects like massive asteroids might occur with terrible frequency, not once in about fifty million years as they do at present, but at least once every hundred thousand. This would make it extremely difficult for a civilisation to evolve, and the simple answer is that there might not be one.[24]

The argument goes like this. Long, long ago, soon after the Sun and the planets were formed, the inner solar system was filled with fast-moving comets. Their impacts splattered the surfaces of the planets. The only trace that remains of them on Earth is the oceans, their melted ice.* Atmospheric weathering, vegetation, and the movement of tectonic plates have long destroyed or hidden their impact craters. But on the surfaces of the Moon and Mercury we may still see these craters, some of which were made so many billions of years ago.

In the outer solar system, that is to say beyond the orbit of Mars, there were also many such comets. They constantly collided with one another, gradually accumulating until they formed *their own* planets, worlds made in their entirety of cometary material. These grew ever larger until they became what we now know as the gas giants Jupiter and Saturn. As time went on, the great masses of these worlds flung out comets that formed the Oort Cloud, and fewer and fewer of them remained to strike Earth.

Moreover, this process continues today. Jupiter, in ancient myth, was the chief of the gods, who terrorised mankind from the peak of Mount Olympus by hurling down thunderbolts. But as science has now shown, he protects us from thunderbolts instead of hurling them. Circling the Sun in its orbit every 12 years like a giant natural hoover, Jupiter either hurls comets out of the solar system or else captures them, adding them to its own mass.

Until 1994 this was largely theory. Then we were privileged to see it happen. Comet Shoemaker Levy 9, which had invaded the solar system from the Oort Cloud, came close enough to Jupiter to be captured. It eventually crashed into the planet after being broken by Jupiter's strong gravity into 11 separate pieces. If Jupiter had not been there, this comet would have eventually hit one of the inner planets, probably Earth, because Earth is the largest of them.

No planetary system, in other words, will offer a safe haven for interstellar travellers, either on its comets or on its planets, unless it has a Jupiter. Fortunately, astronomers searching for alien worlds have found plenty of Jupiters. As I explained earlier, *the reason* they have found these huge worlds – and only these – is because the current technical limits of telescope power only allow the detection of very large planets if they are circling alien suns.

---

* Mars almost certainly once had oceans, from the same source, billions of years ago when it was warmer and closer to the Sun. Wide basins on the Martian surface, seen from the orbiting *Viking* spacecraft in the seventies, show today where these oceans lay.

But the fact that they are there shows that Jupiter and Saturn are not local aberrations, but worlds of a kind that may be found anywhere. Their presence indicates that planetary systems with stable and benign environments may prove to be extremely common.[25]

And what, finally, of Rigil Kent? There seems to be every possibility of finding one or more habitable worlds in this system. The most important recent news – of 1998 – is that a planet with approximately 10 times the mass of Jupiter has been detected by the Hubble telescope orbiting Rigil C.[26] As in the case of Lalande, where there is one planet there may be others. In short we have here, at an extremely convenient distance from us as cosmic distances go, three stars *each* with a reasonably high probability of having a planet in its orbit whose conditions might resemble the Earth's. Rigil A and B, as we have seen, are very similar to the Sun, and neither of them has yet been shown to have *close-in* Jupiter-like planets that would disrupt the orbit or prevent the formation of an Earth-sized world at the right distance from its respective star to nourish life.*

It will take several more decades of telescope development before we know for certain whether habitable worlds exist in the Rigil system. (It may be necessary to place such instruments on the Moon, or beyond the orbit of Jupiter, outside the ring of obscuring dust that surrounds the inner solar system.) But because of Rigil's close proximity to the Sun, it is far the most interesting star system to explore. Even if these worlds prove not to exist, Rigil Kent will contain a mass of useful orbiting debris in the form of asteroids and comets which, to a twenty-second- or twenty-third-century civilisation, would present no obstacle to forming the first interstellar colonies.[27]

Yet whatever our descendants choose to do, and whatever worlds they find out in the realm of the Centaur, the story of Rigil Kent is a saga that is going to continue for thousands of

---

* One planet that defies Black's general prediction is the giant world orbiting the Sun-like star 51 Pegasi, 42 light years from Earth. Its distance from its parent sun is an astonishingly small seven million kilometres, one-twentieth of the Earth–Sun distance! It therefore seems most unlikely that a habitable world could exist around this star, since the existing close-in giant would absorb the materials needed for its creation.

years. Whether or not we go to Rigil, it seems certain that Rigil will come to us. All star systems move relative to one another, and the distance between our solar system and the Rigil stars is shrinking at a rate of about 20 kilometres per second. This means that within about 50,000 years – unless their trajectory is perturbed by the mass of another nearby star – they will come unpleasantly close to us. The orbits of up to 200,000 comets that lie at the edge of our solar system are likely to be disrupted by the gravitational tides of this close approach of a stellar neighbour, many of them turning sunwards and some raining down destructively on Earth.[28]

It will be one more episode that shows how the future destinies of man and the three stars of Rigil Kent are closely intertwined.

Chapter 6

The Sixteen Speeds

When the Man and the Dog came back from hunting the Man said, 'What is Wild Horse doing here?' And the Woman said, 'His name is not Wild Horse any more, but the First Servant, because he will carry us from place to place for always and always and always. Ride on his back when you go hunting.'

Rudyard Kipling, *Just So Stories*

Stand not upon the order of your going,
But go at once.

William Shakespeare, *Macbeth*

There are many ways to measure human progress through the ages, but the one that is most relevant to any prophecy of interstellar travel is the history of speed.

How fast is it possible to travel? From our earliest history as intelligent beings we have been developing our ability to travel faster; not gradually but by sudden bounds. One can draw up a chart of these achievements. In general order of magnitude, there are 16 possible speeds, of which 14 have either been already attained or will one day be attainable. Interestingly, each speed is approximately – very approximately – three times faster than the previous one:[1]

Speed	*Average k.p.h.*
1. Baby crawl	3
2. Walking	10
3. Running	30

4. Galloping horse	50
5. High-speed train	250
6. Jumbo jet	930
7. Concorde	2,250
8. Hypersonic spy plane	11,000
9. Space shuttle	28,000
10. Escape from Earth orbit	40,000
11. Interplanetary	80,000
12. Escape from solar system	150,000
13. Significant per cent of light	80 million
14. Relativistic	750 million
15. Light	1.1 billion
16. Faster than light	Difficult to imagine

For the first 95 or so millennia of human history, people could only attain Speed 3.* With the mastery of the horse about 5,000 years ago, which the French naturalist Compte de Buffon called 'the proudest conquest of man', Speed 4 became commonplace. In the twentieth century, and *only then*, we conquered Speeds 5 to 12.† Robotic visits to the planets have made Speed 11 commonplace. Speed 12 has been reached by the four robotic spacecraft, *Pioneer 10* and *Pioneer 11*, and *Voyager 1* and *Voyager 2*, which, launched in the seventies, have visited all the planets except Pluto and have left the solar system. (Humans – as opposed to machines – have travelled somewhat more slowly. The 21 astronauts who flew to the Moon are the only ones so far to reach Speed 10.)

In a sense, we have been exceeding Speed 4 for many centuries with the use of carrier pigeons. Able to fly at some 90 k.p.h., they were in constant use for sending messages until the invention of radio. In the thirteenth century, Genghis Khan used them to

---

* The greatest known triumph of Speed 3 came in 490 BC after the Battle of Marathon (the name given ever after to long running races), when the Athenians decisively defeated the Persians. The messenger Philippides ran all the way from Athens to Sparta, and back to Athens again, to give the news, a distance of about 240 kilometres in two days. 'We are victorious!' he gasped, and then dropped dead.

† In the nineteenth century we were *on the verge* of attaining Speed 5 but never quite did so. Lift cages in mines reached speeds of 70 k.p.h., and in 1893 the Empire State Express broke records by achieving 180 k.p.h. on the New York central line.

communicate with the extremities of his empire, which extended from China to the Black Sea. In more recent times, news of the defeat of Napoleon at Waterloo in 1815 reached London within a day by means of a pigeon; and in 1849, Baron Paul Julius von Reuter, founder of the famous news agency, started using them to send information between the stock exchanges of Brussels and Aachen.

But we are concerned with much higher speeds, namely with Speed 14 (relativistic speed), almost certainly the maximum that humans will ever attain. (I am confident in saying 'almost certainly', since to go faster would violate physical law. Only light and TV and radio signals travel at Speed 15, and nothing, as far as we know, goes at Speed 16.) 'Relativistic' means that those travelling at Speed 14 will be affected by Einstein's special theory of relativity, and that time on board the ship will run more slowly than on Earth. As I explain later, it will thus be possible to fly to another star and its planets taking half the time for the journey as the same time is measured on Earth. This will not only make it possible for people to travel to the stars in reasonable voyage times, but, as will be seen in Chapter 8, 'The Ascent of Rip van Winkle', will also enable them to make a financial profit out of doing so.

But long before such a stage is reached, it must be recognised that there is one formidable obstacle to travelling faster – to moving from one of the speeds to a higher one. It is *psychological* rather than technical. Throughout history, people have had a strange tendency to believe that to move up a level would be not merely inherently impossible, but dangerous and somehow impious.

The 'If-God-had-meant-us-to-fly-he-would-have-given-us-wings' syndrome dates back to pre-history. This is clearly apparent from the Bible, which, among many other things, is the history of a people who showed an extraordinary reluctance to jump from Speed 3 to Speed 4. As one scholar remarks, 'from early Israel through the monarchy, the Old Testament exhibits a prejudice against the use of the horse and chariot whose base is not entirely apparent'.[2] The favoured method of travel was the much slower ass or donkey. The prophet Isaiah thundered: 'Woe to them that . . . stay on horses and trust in chariots . . . They look not unto the Holy One of Israel, neither seek the Lord.'[3] The laws of the Patriarchs specifically prohibited the kings from acquiring both a large number of wives and a large number of horses, and, indeed, it was considered one of the sins of King Solomon that he kept stalls for thousands of horses at his stables at Meggido.[4] For this prohibition no reason was given. Horses and chariots were for such imperialist foreign enemies as Egyptians and Assyrians. Joshua, instead of making use of the horses and chariots he captured from the Canaanites, had the former hamstrung and the latter burned. David did the same to the horses and chariots that he captured from the Aramaeans.[5] We may recall the contemptuous offer of the Assyrian officer to give the Judahites 2,000 horses 'if thou be able to set riders upon them'.[6]

Horses, indeed, were so vulnerable to climate change that they might have become extinct at the end of the last Ice Age 11,000 years ago were it not for a few enthusiasts who domesticated them. Although they evolved over millions of years in North America, the North American Indians had no use for them until Christopher Columbus reintroduced them in 1494. Until then, the Indians merely dismissed them as 'big dogs'.[7]

European peoples in ancient times, by contrast, *dreamed* of horses that were capable of travelling at unlimited speeds. Two of the earliest stellar constellations, Pegasus and Equuleus, were named after horses, not to mention Sagittarius, Centaurus and Monoceros, which were named after centaurs. All of these fabulous steeds of the sky had *wings*, since their inventors imagined them racing across the sky at unimaginable velocities, prehistoric visions, perhaps, of future starships. In the *Iliad*, Homer tells us how the Greek hero Bellerophon rode the great horse Pegasus across the sky to kill the monstrous Chimera. He strafed the creature with arrows launched like space missiles. His enemy the Chimera was a dreadful hybrid, like an alien monster, 'a thing of immortal make, not human, lion-fronted and snake behind, a goat in the middle, and snorting out the breath of the terrible flame of bright fire'.[8]

Even when the use of horses in peace and war became universal, they were employed inefficiently for many ages. The Roman cavalry did not even have stirrups, a failing that brought them many defeats by the Huns who, riding at full gallop and standing in their stirrups, shot heavy arrows 'with unerring aim and irresistible force'.[9] The invention of stirrups must have doubled the distance that a horseman could ride in a day without tiring. Even after the Huns invaded Europe the lesson was not fully learned. Mounted troops in the early Middle Ages showed no conception of the extra speed and force that stirruped riders could attain. In the words of one historian:

> The Anglo-Saxons used the stirrup, but did not comprehend it; and for this they paid a fearful price. While semi-feudal relationships and institutions had long been scattered thickly over the civilised world, it was the Franks alone – presumably led by the genius of Charles Martel [the victorious Frankish king] – who fully grasped the possibilities inherent

> in the stirrup and created in terms of it a new type of warfare supported by a novel structure of society which we call feudalism.[10]

By the sixteenth century, long after use of the stirrup had been thoroughly grasped, travel times were still intolerably slow by modern standards. William Manchester, in his book *A World Lit Only by Fire*, has recorded the time it took to travel between Venice – Venice then being the centre of a declining but still huge trading empire – and various places in Europe and the Middle East.[11] The average journey times of those days, taking into account all the inevitable delays, were little more than Speed 1, 'Baby Crawl':

Venice to Brussels	16 days
Venice to London	27 days
Venice to Constantinople	37 days
Venice to Lisbon	46 days
Venice to Damascus	80 days

Many centuries later, mankind was readying itself for the jump to Speed 5. But first a digression. Ocean-crossing steamships strictly speaking belong somewhere between Speeds 2 and 3. But they should perhaps be in a category of their own because of their ability to carry huge numbers of people across vast distances in both peace and war. They played an important part in spreading people across the globe – as starships will one day do for the galaxy.

A leading supporter of steamships was the British inventor Henry Bell. As early as 1803, he tried in vain to persuade the Lords of the Admiralty of their obvious advantages: they could move against the wind and tides, move anywhere, in any circumstances, provided only that there was sufficient depth of water on which to float.* His only supporter among the Naval Lords was

* At about this time, Bell's American rival Robert Fulton was trying to persuade Napoleon that steamships would enable him to invade Britain. In Sir Arthur Conan Doyle's historical novel *Uncle Bernac*, the Emperor listens to him impatiently and then exclaims: 'What, sir, you would make a ship sail against the wind and currents by lighting a bonfire under her decks? I pray you excuse me. I have no time to listen to such nonsense.'

Nelson. 'If you do not adopt Mr Bell's scheme,' said the future victor of Trafalgar, 'other nations will, and in the end vex every vein of this Empire. It will succeed and you should encourage Mr Bell.' But their lordships decided that Bell's scheme had no value, leaving a historian of transport to exclaim, in words that ring just as true today as then: 'Why is it that official bodies are nearly always the last to be convinced of the worth of improvements and the inevitability of progress?'[12]

The change to Speed 5 did not come quickly. The first trains – like those that ran on the famous Stockton and Darlington Railway which opened in 1825 – were even slower than horses, and many people felt they should remain that way, if not be prohibited altogether.* One property owner hired gunmen to patrol his land with orders to shoot any trespassing railway surveyor.[13] An editorial in the London *Quarterly Review*, jeering at a proposal to build a railway from central London to Woolwich, declared the project pointless because river boats would get there faster – even though the river twisted while the railway would run straight. 'Old Father Thames against it for any sum!' the newspaper said.[14] But as trains started to go faster, some people objected to their speed. In 1829, when Martin Van Buren was Governor of New York, he wrote an angry letter of complaint to President Andrew Jackson:

> As you well know, Mr President, 'railroad' carriages are pulled at the enormous speed of 15 m.p.h. [24 k.p.h.], by 'engines' which in addition to endangering life and limb of passengers, roar and snort their way through the countryside, setting fire to the crops, scaring the livestock and frightening women and children. The Almighty never intended that people should travel at such breakneck speed.[15]

And so to aircraft. Even in 1903, the year of the first powered flights by the Wright brothers at Kitty Hawk, North Carolina,

---

* The trains were so slow that coach-and-horse drivers were allowed to use the rail track when the former were not running. The railway company even paid coach drivers to use them, giving them £25 for horse and harness. But it took no chances of their obstructing the trains. One clause in such a contract given to a certain Thomas Close stated: 'The first time he is seen intoxicated he will be dismissed and the sum due to him will be forfeited.'

their prospects of success were dismissed with contempt by transport experts. Professor Simon Newcomb, a distinguished astronomer at Johns Hopkins University, gave a definitive scientific proof that heavier-than-air flying machines were 'utterly impossible'. His mistake, obvious in retrospect, was failing to understand the idea of the 'aerofoil' – the principle by which air flows faster *over the top* of a flying wing than underneath, thus creating a reduced pressure above the wing, causing it to rise.[16]

It is less generally appreciated that even in the late twenties, after aeroplanes had performed well in battles of the First World War, many prophets could see no future in them. One authoritative writer, Marion W. Acworth, declared in a book published in 1927 that they would bring only delay, discomfort and the threat of war, and that they would therefore soon be abandoned.* Today, her predictions make ludicrous reading:

> The number of our countrymen wishing and willing to fly [from Britain] to India, even with State assistance towards the fare, in a series of 'hops' over desolate mountains and trackless deserts, rather than by a restful and refreshing sea voyage, must be few indeed; for aerial transport is admittedly comfortless, more dangerous, more unreliable and more expensive than other means of travel . . . Aeroplanes can never be made to pay in peace as passenger or freight carriers, and in war they have proved themselves to be unreliable, ineffective and unprofitable, no matter how brave the pilots or spectacular their exploits. While prodigal of life and treasure, aerial warfare has only succeeded in sowing mistrust and enmity, breeding fear, encouraging frightfulness, provoking thereby a manifest return to barbarism . . .
>
> It is interesting to examine the recent flight of four RAF machines from Cairo to the Cape of Good Hope [South Africa] and from the Cape to London. Reflection shows the absurd disproportion between the effort and the useful attainment. At very great cost, four machines aggregating

* The great aviation pioneer Frederick Lanchester (1868–1946) once received an oral message to this effect from the Sea Lords of the British Admiralty – that they were opposed to his research into aircraft. He merely smiled. 'But Mr Lanchester,' said a horrified official, 'you do not seem to know what a Sea Lord is.' 'Oh,' said the inventor, 'I suppose it is one between a B-Lord and a D-Lord.'

> 1,800 horse power conveyed eight persons a distance of 1,400 miles [2,200 kilometres] in 114 days, giving an average continuous speed of 5 m.p.h. [8 k.p.h.]. It is stated that the machines kept to a continuous timetable. With the same horse power, but at comparatively trifling cost, two tramp steamers running also to scheduled time could have conveyed 4,600 tons of cargo a similar distance in half the time.[17]

An Army officer added this further absurdity:

> Of the free air rushing past, the passengers feel almost nothing. They thus lose much of the exhilaration of high-speed flight, and they miss the vitalising, clarifying air-bath which is given to the pilot and passenger in an open machine. If air travel is ever to be as good for the health as sea travel, it will be necessary on air liners of the future to provide a number of open seats in which be-goggled, hide-wrapped passengers can have fresh air poured into their lungs.[18]

But despite all this scepticism, speeds have continued to increase. They reached, by the end of the twentieth century, what statisticians call an 'exponential curve', an upward curve that continuously accelerates and shows no signs of slowing. In the graph opposite the abbreviation 'psol' means percentage of the speed of light. The graph may be plotted as shown.

Drawn in 1953, this graph has proved uncannily accurate.[19] It more or less 'predicts' that artificial satellites would fly a few years later, achieving Speed 9. (So they did. Russia's *Sputnik*, the first of thousands of such craft, flew in 1957.) Moreover the graph predicts that Speed 12, escape from the solar system, would be attained at about the turn of the century. Is this too optimistic? Perhaps only a little. As related earlier, four deep-space probes, two *Pioneers* and two *Voyagers*, having left the realm of the planets behind them and entered interstellar space, have accelerated to about 60,000 k.p.h.

It is of course only chance that this pre-Space Age graph has proved so accurate. It might easily not have done so. *Pioneer* and *Voyager* might have malfunctioned or fallen victim to budget cuts. But the graph has turned out to be accurate, and we are entitled to draw conclusions from it. It tells us – and bear in mind that it is following a trend in speed records going back before 1800 – that

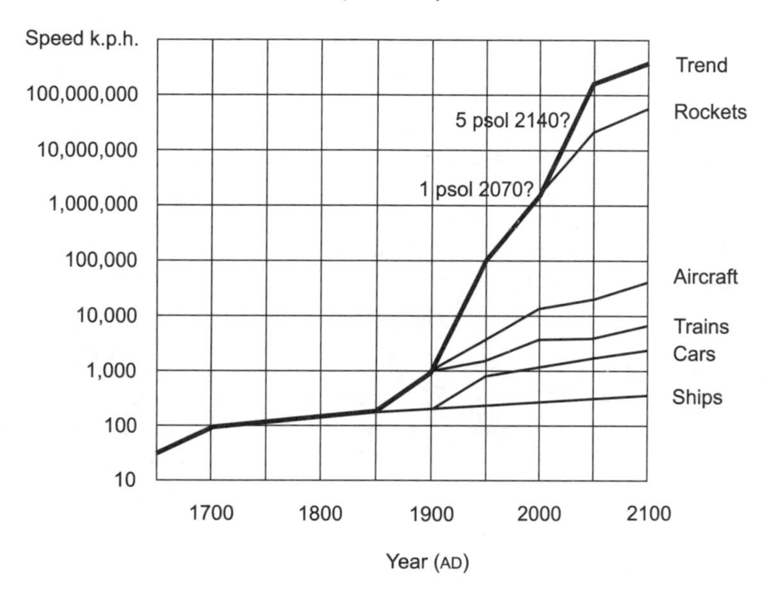

the achievement of Speed 13, a significant percentage of the speed of light, will come between 2070 and 2140 when we reach velocities of between 1 and 5 per cent of the speed of light. If this trend in rising speeds continues, we should reach the all-important Speed 14, relativistic speed or about 90 per cent of the speed of light, which will enable people to fly to the nearer stars, at some point in the twenty-third century.

But will we really do so? For that is all it is – a trend. As the economist Sir Alexander Cairncross warned in a famous poem:

> A trend is a trend is a trend.
> But the question is: will it bend?
> Will it alter its course, through some unforeseen force,
> And come to a premature end?[20]

Will the trend of rising speeds 'bend' and 'come to a premature end'? Ultimately, it must do. As James Strong points out in his book *Flight to the Stars*:

> An exponential growth cannot go on doubling for ever. Long before the snowball reaches the valley below, it crosses the snowline and ceases to grow. Long before the number of people entering bureaucratic employment exceeds the entire population, sheer physical necessity will force a change in policy. An exponential curve is only valid in practice *up to a point*. At some moment in its life, saturation takes place and the rate of change must slow down, level off, or even reverse.[21]

It will not reverse in the case of speed, but it must level off. For the speed of light, Speed 15, 1.1 billion k.p.h., appears to be an absolute barrier to further acceleration. It is forbidden by physical law. Only light beams and radio signals travel at this speed. No machine can reach it.

But from the point of view of interstellar travel, this need not matter. It would indeed be more convenient if it were otherwise, if we could travel to the stars at any speed we wanted, commuting, for example, to Proxima Centauri to do our Christmas shopping. I explore this possibility in a later chapter. But it remains just that – only a possibility, whose faint hopes of coming true are still hidden in the depths of physics. It will be more realistic at present to bank on sub-light speeds.

Nonetheless there is an all-important consolation. We will not have to waste time making inordinately long voyages, because at Speed 14 the passage of time itself will slow to suit our convenience.

Chapter 7

The Incredible Shrinking Calendar

Time travels in divers places with divers persons. I'll tell you who Time ambles withal, who Time trots withal, who Time gallops withal, and who he stands still withal,

William Shakespeare, *As You Like It*

If we planted a living organism in a box . . . one could arrange that the organism, after an arbitrarily lengthy flight, could be returned to its original spot in a scarcely altered condition, while other organisms, which had remained in their original positions, had long since given way to new generations. In the moving box, the length of time of the journey was a mere instant, provided the motion took place with approximately the speed of light.

Albert Einstein, 1905

The difficult prose of the second quotation is typical of the language of physics, but try to make sense of it. It needs only two tiny changes for its meaning to become obvious. Replace the word 'box' with 'spaceship', and 'organism' with 'astronaut'. We then have a clear statement of one of the most shattering discoveries of this or any other century: that the faster human beings travel through space, the more slowly they age.

This will not of course occur at any speed at which we are familiar. Only in space vehicles with yet barely imaginable velocities will such changes take place. Note how Einstein is careful to say: 'provided the motion took place with approximately the speed of light'. Not even at the fastest speeds at which we travel,

on a flight in Concorde or circling the world every 80 minutes in a space shuttle, will the passengers encounter any change in the rate of passage of time – any 'time dilation' – which their wristwatches could possibly measure.*

---

* This is not true of atomic clocks, which are vastly more accurate than wristwatches. In 1971, two physicists, Joseph Hafele and Richard Keating, did an experiment that confirmed Einstein's prediction. They took two atomic clocks in a jet aircraft that travelled round the world, having first synchronised them with two identical clocks that remained on the ground. When the clocks were compared after the flight, it was found that the *airborne* instruments, during the flight, had been running at just under 60 billionths of a second more slowly than those on the ground.

To put this matter numerically, here is a table to show the rate at which time slows depending on speed[1] (I again use the term psol as a percentage of the speed of light):

Speed of ship (psol)	*Duration of ship hour (minutes)*
0	60.00
10	59.52
20	58.70
30	57.20
40	55.00
50	52.10
60	48.10
70	42.85
80	36.00
90	26.18
92	23.52
95	18.71
99	8.53
99.9	2.78
99.997	1.17
100	zero

The predictions given in these figures, derived from Einstein's special theory of relativity of 1905, constitute one of the principal reasons why travel to the stars *will* be possible, and also, as will be seen later, economic and profitable. This phenomenon of time dilation came as a tremendous intellectual shock when Einstein proposed it at the start of the century, and there are still many – and there will be many in the future, when it actually happens – who refuse to accept it. I believe that the prediction of James Strong, in his book *Flight to the Stars*, will prove as true two centuries from now as it was when he wrote in 1965:

> Of all the strange and seemingly irrational effects displayed by interstellar flight, none has received more attention, and none is more difficult to explain, than the peculiar anomaly of time dilation. The majority of people cannot conceive of time being any different to star travellers than to those who will be left behind on Earth. It refuses to make sense to read that a starship which took twenty years to travel to [the star] Procyon and back should have a crew who quite

> genuinely and sincerely insist that the double journey only took three years.[2]

The most difficult problem facing what Strong calls the 'majority of people' is not knowing precisely what time *is*. The story is told of the Russian poet Samuel Marshak who, visiting London some time before 1914, found himself without a watch. In bad English, he asked a man in the street: 'Please, what is time?' The man looked surprised and replied: 'But that's a philosophical question. Why ask me?'[3]

Even scientists did not feel that they had any precise ideas about the nature of time until the age of Isaac Newton, who defined it – wrongly – as a process that 'flowed', always at the same pace, no matter where one was or what one was doing. As Newton wrote authoritatively: 'Absolute, true and mathematical time, of itself, and from its own nature, flows equably and without relation to anything external.'[4] Only Newton's contemporary Gottfried Leibniz (with whom he constantly quarrelled) dissented from this view, arguing that time was not absolute but *relative*, and that 'instants apart from things are nothing'. But it was Newton who was believed. Leibniz, after all, was talking only as a philosopher; and it was not at all clear what he meant by the obscure statement 'instants apart from things are nothing' – it certainly isn't clear to me. Newton, on the other hand, in so far as was possible in the seventeenth century, was basing his opinion on the observed behaviour of the real universe.

It was not until the twentieth century, after tremendous progress in physical discoveries, that Einstein could demonstrate that, at the extreme velocities depicted in my table on page 93, Newton's view of time breaks down completely.

But a note of caution. Of the greatly differing universes of Newton and Einstein, many people take the view that Newton was 'wrong' and Einstein was 'right'. This is a gross oversimplification. For all practical cases that we are liable to encounter today, Newton's laws of gravity are perfectly accurate. The Moon stays in its orbit because of his First Law, and rockets propel themselves through space because of his Third Law.[5] However Newton's laws break down, and yield to Einstein's, in extreme circumstances such as very strong gravitational fields and speeds close to that of light. It is with these that we are concerned in this chapter.

'Without relation to anything external.' That phrase of Newton's represented the common-sense view of time that prevailed until the end of the nineteenth century. Time, to almost all scientists, did not seem to *be* anything. It was only the rate at which events happened. The idea that if one travelled fast enough one could return and find oneself younger than one's own children would have seemed to them too preposterous for contemplation, akin to philosophy and jokes like that of the Mad Hatter in *Alice in Wonderland*, who claimed that where he lived it was always six o'clock. As the cosmologist George Gamow recalled in 1947: 'So strong was the belief in the absolute correctness of Newton's statement about the nature of time that it has often been held by philosophers as given *a priori*, and for three centuries no scientist even thought about the possibility of doubting it.'[6]

Then, in 1887, came an experiment that was to have a shattering effect on our view of the nature of time. The conclusions to be drawn from it were so bizarre that at first it was believed that the equipment being used must be faulty. But when repeated again and again, with ever more sophisticated and accurate instruments, it dawned on scientists that a totally new view of the universe was at hand.

Two American physicists, Albert Michelson – already famous for his discovery that the speed of light was almost exactly 300,000 kilometres per second – and his colleague Edward Morley, were trying to discover evidence of the 'luminiferous ether'.

Belief in the existence of the 'ether' – or *aether* – not to be confused with the chemical of the same name, had been widespread for many centuries. It was the invisible substance that was believed to fill all space. The very name 'ether' became a byword for mystery, a thing that existed but that remained indefinable and unseen. As Disraeli said of a character in his novel *Tancred*, 'her ethereal nature seemed to shrink from coarse reality'. Ether had to be real, so the reasoning went, because starlight had to travel *through* something, just as a current travels through water. As the Dutch philosopher Benedict Spinoza said in the seventeenth century, likening the supposedly all-pervasive ether to God, 'nature abhors a vacuum'.[7]

Plato called the ether a 'material substance of a more subtle kind than visible bodies, supposed to exist in those parts of space that are apparently empty'.[8] The nineteenth-century physicist James Clerk Maxwell, in the belief that light's speed was finite because it was obstructed by this same ether, remarked that this substance had been 'invented for planets to swim in'.[9]

Maxwell's calculations, and Michelson's measurements, had shown that light had a fixed speed. But fixed relative to what? It could only be the ether. It was this consideration that led Michelson and Morley to attempt their experiment in 1887 to find definite proof of the ether's existence. Yet instead of finding it, their investigation was to lead to the gradual discovery that all this talk about an ether was nonsense, as was Newton's belief in absolute time, and that the universe was a far stranger place than anyone had dreamed.

To prove the existence of the ether, the two men set up a slab of stone in Morley's basement laboratory in Cleveland, Ohio. They were a contrasting pair: Michelson was erect and dapper, maintaining the appearance of the Naval officer he once had been, while Morley was burly and unkempt, with his hair down to his shoulders and a great bristling red moustache that straggled almost to his ears.[10] Their slab rested on liquid mercury to eliminate vibrations. It was about two metres long and a third of a metre thick. Their plan was to use light beams to measure the speed of the Earth's passage *through* the ether. Because of Earth's movement through space, they reasoned that light waves would move slightly more slowly along the slab in the direction of the planet's motion through the ether than they would when going away from it – just as a hiker on a windy day feels the wind against his face vary in strength according to which direction he is walking.

But to their amazement, they found that light always travelled at the same speed, 299,792 kilometres per second (300,000 kilometres per second is an excellent approximation), irrespective of the Earth's (and the slab's) direction through space.

Or to put this another way, not only was there no sign of an ether, but light seemed always to travel at *exactly* the same speed, whether one was travelling towards its source or away from it.

Neither Michelson nor Morley, nor anyone else at the time, was prepared to accept this result. It seemed so absurd! Bertrand Russell wrote that if the ether existed, it was hard to believe that it would go to such prankish lengths to hide its existence. He likened its behaviour to that of the White Knight in *Alice Through the Looking Glass*, who announced:

> But I was thinking of a plan
> To dye one's whiskers green,
> And always use so large a fan
> That they could not be seen.[11]

For – and this was the fantastic-seeming objection – if the speed of light *was* constant irrespective of the speed of its source, then, as will be seen in a moment, Newton's view of the absoluteness of time would have to be thrown out, a step that no one was yet prepared to take. And so, back to work! The experiment was repeated several times with ever more sensitive instruments, but always with the same negative results. As Martin Gardner says in his *Relativity for the Million*:

> Michelson was astounded and disappointed. This time the astonishment was felt by physicists all over the world. Regardless of how he and Morley turned their apparatus, they found no sign of an ether! Never before in the history of science had the negative results of an experiment been so positive and so shattering. Michelson once more thought his experiment a failure. He never dreamed that this 'failure' would make the experiment one of the most successful, revolutionary experiments in the history of science.[12]

What way out of this confusion could there be? The strangest explanation of all was put forward independently by an Irish physicist, George Fitzgerald, and a Dutchman, Hendrik Lorentz, both of whom afterwards became close colleagues of Einstein. Suppose, they suggested, the results of the experiment were in fact correct, and that clocks were being slowed down very slightly by the Earth's passage through the ether. They would be slowed in just such a way as to make the speed of light always measure 300,000 kilometres per second. Another consequence would be that the ether puts pressure on a moving object, such as the slab, shrinking it very slightly in the direction of motion. Imagine, instead of a slab, a spaceship shaped like a long, thin cigar. As it rushed through space, its shape would distort until it resembled a short, fat cigar. Of its increase in fatness, more in a moment.

But, said the objectors, if these statements were true, why was it still not possible to measure changes in the speed of light as the slab/spaceship changed its direction of motion?* Because, said the physicists, *the dimensions of the measuring instruments were themselves distorted.* How can one measure the length of a shrinking object with a ruler that itself shrinks?

This idea recalled a strange 'thought experiment' put forward in the nineteenth century by the French mathematician Henri Poincaré (first cousin of the famous statesman). Suppose, said Poincaré, that suddenly one night, the entire universe became a thousand times larger. By the 'universe', he meant *everything*, stars, planets, people, atoms, atomic particles, even wavelengths of light. Would it not be possible to detect the change? No! Because every measuring instrument with which you tried to measure it would itself have changed. Nobody in the universe could prove or disprove that the enlargement had taken place. The event would literally be meaningless. The alteration would only be detectable by someone who lived outside the universe and who had some means of peering into it.†

Shrinking length and the slowing of time are not the only changes that would affect a fast-moving spaceship. The ship, and everything in it, would become heavier as well.

As the length of a ship shrinks, its *mass* increases. Its mass, which may be measured as the amount of energy required to accelerate it further, increases in the same proportion as its length is shortening and its passage of time decreasing. When it is shortened to one-tenth of its length when stationary, it is ten times more massive. In other words, it is offering ten times as much resistance to its rocket motors. Its mass continues to increase, ever

---

* This emphasis on spaceships is my own. Physicists in the twenties, when Einstein's discoveries were first debated, had little conception of space vehicles. They thought in terms of 'boxes' or 'bodies', which must have made it much harder for laymen of the time than our generation – when space travel is commonplace in both fiction and reality – to understand what they were talking about. Indeed, one of Einstein's key papers in 1905 is obscurely entitled, giving little idea of its momentous message, 'On the Electrodynamics of Moving Bodies'.

† To make the same point, Martin Gardner, in his *Relativity for the Million*, invents the amusing story of two sailors, Joe and Moe, who are wrecked on a desert island. One day a new giant-sized bottle of Coca-Cola drifts by. Joe turns pale. 'Hey, Moe!' he shouts. 'We've shrunk!'

more steeply, as it approaches the speed of light.[13]

Consider this in terms of an accelerating spaceship. At the comparatively low maximum speeds of which we are capable today – typically Speed 11, 'Interplanetary', about 80,000 k.p.h., or .007 per cent of the speed of light – any changes in time, length or mass are negligible, only measurable by atomic clocks. But at speeds much closer to that of light, the changes become striking. At half the speed of light, a ship's mass would increase by 15 per cent. At 90 per cent, the ship becomes two and a third times more massive. At 99 per cent, it is seven times more massive. At 99.99 per cent, its mass has increased by a factor of 70, and at 99.9999 per cent this factor is 700! Each time two more nines are added to the right of the decimal point, the ship's mass increases by an additional further factor of 10.[14]

The consequences of this are obvious. At the speed of light itself, the ship's mass would become infinite. To put this another way, to accelerate to the speed of light would require an engine *of infinite power*. And since it is self-evident that no such engine can ever exist, it follows that any speed equal to or greater than that of light is an eternal impossibility. It might seem convenient indeed if things were otherwise, if ships could fly at any speeds they liked, hopping from star to star in weeks or days, as the *Enterprise* does in *Star Trek*! But unless there are short cuts through other dimensions (see Chapter 13, 'Faster Than Light?'), this cannot be. In the universe ruled by Einstein's special theory of relativity (which incorporates the calculations of Lorentz and Fitzgerald), there is an absolute and unbreakable speed limit.*

But this does not mean that interstellar travel is impossible or even impracticable. The slowing of time in a fast-moving ship is a definite advantage for the travellers. It means that a crew would be able to complete a journey to the planet of another star that lasted many years as measured by clocks on Earth, but in a

---

* Einstein went much further than Lorentz and Fitzgerald, which is why we call special relativity *his* theory and not theirs. They attributed changes in the ship's dimensions, and the slowing of time inside it, to the effects of the ether. But to Einstein, they are caused by the gravitational field of the whole universe. Einstein, with characteristic intellectual ruthlessness, declared that the reason why Michelson and Morley had found no sign of an ether is because there *is* no ether.

vastly shorter time as measured by the crew. This is not an illusion to be explained by psychologists. It is a physical fact that would be confirmed by instruments on the ship even if there were no people on board. There is not one, absolute rate of the flow of time in the universe, as Newton believed, but an infinite variety of local times that depend on the speed through space of the people measuring them.

Can we detect such 'local times', not in some far-off technological future, but here and now? Indeed we can. Recently the University of Utah's cosmic ray detector at Dugway, near Salt Lake City, discovered the fastest stream of cosmic rays ever found. Known by amazed observers as the 'Oh-my-God particles', they were crashing into the Earth's atmosphere with 10 trillion times more energy than any created in a man-made atom-smashing machine. They were arriving at the fantastic speed of 99.99999999999999999999999999951 per cent of the speed of light.[15]

Now although it seems very unlikely that any civilisation could build a ship that travelled as fast as this, imagine the consequences if they did. The ship's crew would reach very distant destinations, as measured by *their* clocks, in astonishingly short periods of time. They would take only 4.3 thousandths of a second to reach the nearest star to the Sun, Alpha Centauri, a distance of 4.3 light years. A journey to the centre of our Milky Way galaxy, 30,000 light years away, would take them 3.2 seconds.* A voyage to the Andromeda galaxy, 2.2 million light years away, would last 3.5 minutes. And the Virgo cluster of galaxies, some 40 million light years away, would be reached in one and a half hours. A trip of 2,500 million light years to the brightest quasar, 3C273, would last three days, and they would reach the edge of the universe, an estimated distance of 17 billion light years, within 19 days.[16] This may all sound very hard to believe, but given such speeds, these journey times are correct.

---

* It is amusing to note that the fictitious Starship *Enterprise* cannot begin to match these voyage times, despite having 'abolished' special relativity with its prohibition of faster-than-light travel. Its fastest speed is Warp Nine, or 1,516 times the speed of light. But this would take the ship a full 20 years to reach the galactic centre, no matter how this time was measured. Gene Rodenberry, founder of the series, once said, inexplicably, that it would be 'too dangerous' to introduce real science. It seems much more dangerous not to!

Indeed Einstein's universe sounds so bizarre that it recalls the couplets:

> Nature and Nature's laws lay hid in night:
> God said: 'Let Newton be!' and all was light.
>
> It did not last. The Devil, howling: 'Ho!
> Let Einstein be!' restored the status quo.[17]

The Devil, in this case, has been slow to influence our general thinking. At school (and that was a long time after 1905!) I remember singing a hymn containing two lines that showed the long-lasting authority of Newton:

> Time, like an ever-rolling stream,
> Bears all its sons away.[18]

This statement is inaccurate. Time will not always bear *all* its sons away, only those, like everyone today, who travel comparatively slowly. And if – impossibly – one was able to ride on a light beam, to travel at the speed of light itself, one would never be 'borne away'. One would literally live for eternity, witnessing that distant epoch when life will be no more because there will be no atoms.

But to return to a future that is much closer to us, of what practical use are the seemingly arcane calculations of special relativity? What profit would there be, financial or otherwise, in travelling to a planet circling the star Tau Ceti, 12 light years away, at 92 per cent of the speed of light – much slower and more practicable for a man-made ship than the velocity of the Oh-my-God particles – in a journey of just over *three* years, while the journey time would be measured *on Earth* as taking almost 13 years?*

The answer is that there would be a very great profit, and it *would* be financial.

---

* 92 per cent of the speed of light is considered by some experts an ideal compromise between the demand for speed and the constraints of energy. I use it throughout this book as a proposed cruising velocity.

Chapter 8

The Ascent of Rip van Winkle

This thing all things devours:
Birds, beasts, trees, flowers;
Gnaws iron, bites steel;
Grinds hard stones to meal;
Slays king, ruins town,
And beats high mountain down.

J.R.R. Tolkien, *The Hobbit*

What is 'this thing'? The above verse riddle is the ultimate challenge issued by the evil Gollum to the hobbit Bilbo when the latter is trapped in the goblins' caves. Bilbo is baffled. He sits in the dark 'thinking of all the horrible names of all the giants and ogres he had ever heard of in tales, but none of them had done all these things'. He at last cried out: 'Give me time! Time! Time!', and time of course is the answer.

It is also the answer to another important riddle: why people will find it profitable to undertake voyages between the stars. So far I have confined myself to describing the technical arguments for them. But without a motive the technical arguments are meaningless. The big question is *why?* Who will pay for starships, and *why* should they pay for them?

Before explaining why time is itself a profitable interstellar commodity, let us see how writers on the subject, whether of fiction or non-fiction, have addressed the question of *why?* They seldom address it at all! They generally assume that once starships have been built, with the capability to achieve a large percentage of the speed of light, a significant proportion of mankind will simply 'take off' and roam the galaxy. Such travellers, according to this optimistic view, will comprise merchants, brigands, military adventurers, scientists, artists, tourists, writers, and anyone else who can raise the fare to experience the wonders of unimagined and unimaginable realms. And those few who do address the question somewhat carelessly assume that trade will be the dominant motive. But trade in what? At this point they become vague. Surely there will be something to trade in! In an earlier book I went along with this enthusiasm, and quoted James Strong on the subject:

> Wine from the slopes of Earth's vineyards may change hands in token for curios from the hot stars of the Trapezium, rare earths exchanged for jewels, or drugs from glittering Polaris, silks and furs from Arcturus for insect-pets from far-off Wezen. There will be an upsurge of merchant adventuring never before witnessed if interstellar traffic gets a hold on men's minds.[1]

Strong, so wise on technical aspects of interstellar travel, seems considerably less so on this matter. I am not sure that the merchants whose activities he imagines would stay solvent for very long. The real-life merchants who would like to profit from trade in these supposed treasures are hard-headed people who will certainly ask tough questions before fitting out an expensive trading expedition.

How, for example, would they *know* there are 'curios' to be had for the taking on planets circling the stars of the Trapezium that are not only valuable enough to be exchanged for crates of French or Californian wine, but so valuable that their worth will survive the decades needed to ship them home? What 'jewels' and 'drugs' are likely to exist on the worlds of Polaris that cannot be manufactured or found on Earth? The most exquisite silks are made in China, India, Japan, and Thailand. Why should one go to Arcturus to seek better samples? The star Wezen is nearly 2,000 light years away. How would they know that its planets harbour potentially valuable 'insect-pets', and even if they do, how could they be sure that they will not have lethal bites or that they will not breed us out of hearth and home? It seems a long way to travel just to carry out an insect-pet trade feasibility study.

Imagine, even in the twenty-second century, seeking a loan from one's bank manager to finance such vague schemes as these!

Interstellar travel, being a large and expensive industry, will of course generate huge sums of money in the form of construction contracts, sub-contracts, and sub-sub-contracts to the nth degree. But that is not to say that it will be economic. The aviation industry, the closest modern analogy to such future activities, provides Boeing and Airbus with enormous revenues, but, to make an obvious point: it is not these aircraft manufacturers that make the industry flourish, it is the passengers. If no one was willing to buy an air ticket, all the jumbo jets in the world and their expensive fittings would be worthless white elephants. As the space writer Warren Salomon eloquently points out in his essay on 'The Economics of Interstellar Transport':

> Let's be clear about what's involved in this question. I'm not talking about whether anyone can profit from ship building, or supplying equipment for the crew, or providing fuel, or any of a thousand other peripheral activities. If someone is willing to pay for interstellar voyages, there will of necessity be a demand for ships, equipment and fuel; and businesses will naturally spring up to fill those needs. What I'm getting at is a more fundamental issue – what makes the *whole venture* economically worth the bother. Where will the funds come from to pay for interstellar voyages?
>
> Science fiction literature is filled with endless blather about the 'sense of wonder', mankind's love of adventure, and our insatiable quest for knowledge. These motives exist, and no doubt some ships will be launched for those reasons. But what will keep the activity going on a continuing basis? Why will fleets of ships be built and launched, decade after decade, generation after generation, century after century? Just for the sheer joy of it? I think not.[2]

The same objections seem to apply to the trade of *any kind* of interstellar goods. Will we want ore? Ores can be had in plenty from mining our own asteroids or the outer moons of our solar system. Hydrocarbons? Hydrocarbons exist in abundance in the atmospheres of our giant gas planets, Jupiter, Saturn, Uranus, and Neptune. What of the metals platinum and palladium, rare and expensive on Earth but essential ingredients for catalytic converters that remove noxious gases from motor car exhausts and that will soon be compulsory for new cars in many countries? They exist in profusion, and will be available much more cheaply, in many of the asteroids.[3]

Is there perhaps some analogy to present-day intercontinental trade on Earth? What of the Japanese, who import vast amounts of raw materials, make them into cars, and sell them to Europe and the United States? No, there is no parallel here, for completed Japanese cars take only a few weeks to be shipped across the oceans to their points of sale, while Toyotas made on a planet circling Betelgeuse would take *decades* to reach Earth.[4] It is hard to believe that automobile fashions and technologies would not have changed while these cargoes of cars cruised through interstellar space without making any profit!

In final desperation we might turn to the causes of the great movements of oceanic exploration that began in the fifteenth century. Here we are not dealing with anything as fundamental as industrial raw materials but with the quest for food and other dinner table consumables. Several science fiction writers have explored this possibility, so let us consider it.* Why did such men as Columbus and Magellan set forth across unknown oceans? Magellan's biographer Stefan Zweig gives this answer:

> The quest for spices began it. From the days when the Romans, in their journeys and their wars, acquired a taste for the hot or aromatic, the pungent or intoxicating dietetic adjuvants of the East, the western world found it impossible to get on without a supply of Indian spices in cellar and storeroom. Lacking spices, the food of Northern Europe was unspeakably monotonous and insipid, and thus it continued far on into the Middle Ages. Centuries were to elapse before the fruits, tubers, and other products which now seem commonplace were to be used or acclimatised in Europe. Potatoes, tomatoes and maize were unknown. There were no lemons to prepare acid drinks, there was no sugar for

---

* The most famous of these SF novels is surely Frank Herbert's *Dune*, with its fabulous spice that is available only on the desert planet Arrakis. But it could be a mistake to count on the possibility that there will be *one* vitally needed commodity that will be found only on *one* planet in the galaxy!

> sweetening, the cheering tea and coffee were still lacking; even at the tables of the rich and powerful there was naught to relieve the sameness of perpetual gluttony – until, wonderful to relate, it was found that a touch of spice from the Orient, a dash of pepper, a minute addition of grated nutmeg, the mingling of a little ginger or cinnamon with the coarsest of dishes, would give an unwonted and wholesome stimulus to the jaded palate.[5]

But this motive will have no power among the stars. For it was not distance that prevented northern Europeans from obtaining their spices and fruits from the East. It was greedy middle-men who sat on the trading routes, taking their cuts, until prices rose so intolerably that very rich Europeans became known as 'pepper sacks'. In short, what drove Europeans to seek other trade routes to the East was not *natural* difficulty – as is the case with interstellar distances – but *human* obstruction; and that can be eliminated by negotiation or war, or in this case by finding a short cut. Once again, it appears that every commodity which human beings living in the solar system will ever want will be obtainable *inside* the solar system, from sources millions of times closer than the stars.

Every commodity? There is, as I suggested above, perhaps one that can only be obtained from interstellar travel, although it is not one that can be eaten, drunk, or turned into industrial machinery. It is another idea that I owe to Warren Salomon. It is *time*, the answer to Gollum's riddle.

One of the most famous American legends is the story of Rip van Winkle, an idle, thriftless fellow who in the late eighteenth century wandered one day with his dog and gun to the peak of a magic mountain in the Catskills of New York. There he met some strange people and, taking a drink from them, at once fell asleep. When he awoke, his dog had vanished and his gun was covered with rust. He went down the mountain and found no one in his village that he knew.* It had aged 20 years while he had slept but one night. He had gone to sleep a subject of King George III and

* His new situation came as a relief rather than a shock since his tyrannical wife, who had always blamed him with bitter vituperation for loafing around ale-houses instead of doing some work, was long in her grave.

awoken as a citizen of the United States.[6]

Now suppose that Rip had *not* been idle and thriftless but a keen businessman and one who had known in advance about the properties of drinks to be had on the magic mountain. What if he had possessed a considerable amount of money and, before starting his climb, had invested it at compound interest? On his return, instead of being denounced as a British spy, he would have won universal respect because he would have been fabulously rich.

Suppose he had invested $100 at a compound interest rate of 11 per cent (I shall explain in a moment why 11 per cent is a realistic rate). On his return he would have found that his capital had appreciated by *800* per cent. Together with the sum he had invested and which was now repaid, he would be the richer by $900. It would have been quite a profitable trip to the mountain.

Indeed, he would be in a doubly fortunate position. He would only be a day older than when he had first gone to the mountain. What should he best do? Why, it is obvious. He would re-invest his $900 with various local people (spreading the wealth in case one or more of them went bankrupt or absconded during his absence) and proceed once more to the magic peak. On his second return another 20 years later, his $900 would have turned – together with the original capital – into $8,000. He could repeat this exercise perhaps ten times. By the end of the twentieth century, he would be only 12 days older, but his village would be 240 years older. His original $100 would by then have turned into something like $7 *trillion.*[7]

It might be objected that although Rip becomes steadily wealthier, his life is somewhat dull. He divides it between bargaining with merchants, walking up and down mountains, and accepting peculiar drinks from peculiar people. But he does not have to take all the money for himself. He could collect some friends and turn the enterprise into a partnership. When he wearied of ascending the mountain, others could take over the task, leaving him to invest his fortune elsewhere. By this means a web of wealth would be created, its owners spreading it and themselves through time and space.*

* Another useful literary analogy is H.G. Wells's 1898 novel *The Sleeper Awakes*, the story of a man who goes into a coma that lasts 200 years. When he awakes,

Enough of fairy tales. Let us return to the real world. It should be obvious that Einstein's special theory of relativity will turn high-speed interstellar voyages into real-life equivalents of Rip's ascents to the magic mountain. Imagine a merchant adventurer in the twenty-second or twenty-third century who might wish to raise some money that would – ultimately – pay for a voyage to another star many times over. His conversation with a banker might go something like this:

BANKER: Well, sir, I suppose you have come to ask me for a loan.
MERCHANT ADVENTURER: A loan is certainly something I want to discuss.
BANKER (*spreading his hands*): In that case I sincerely hope I shall not be obliged to disappoint you. Money at present is extremely tight. Only last week we had a very tough memorandum on the subject from our board of directors—
MERCHANT ADVENTURER (*smiling*): The tightness of money is a problem I can solve.
BANKER (*looking baffled*): I don't think I quite understand.
MERCHANT ADVENTURER: I said I was here to discuss a loan. But I don't want to borrow money; I want to lend it.
BANKER: To us?
MERCHANT ADVENTURER: Precisely.
BANKER (*smiling cautiously*): That would indeed be very agreeable. But what would be the terms and conditions?
MERCHANT ADVENTURER (*producing a cheque for a huge amount*): Here is the sum that I want to lend you. I want you to invest it for me at compound interest. A rate of 11 per cent annually would be acceptable—
BANKER (*startled*): That is a very high rate of interest!
MERCHANT ADVENTURER: Not if I undertake not to touch the money for 20 years.* Then, when that period has elapsed, I, or my representative, will return to ask for the money back plus all the accumulated interest it has earned.

---

his friends having invested all his money on his behalf at compound interest, he finds himself the richest person in the world – much to the annoyance of the self-appointed Trustees, who have long wielded political power in his name.

* Various financial experts have advised me that in these circumstances, where the money was not going to be touched for many years, it would be realistic to expect a compound rate of interest of 11 per cent.

BANKER (*still looking cautious*): I see. But beside the usual banking fees, what is in it for us? How will this transaction loosen the tightness of money?

MERCHANT ADVENTURER (*gesturing triumphantly towards the cheque*): For a full 20 years you will have the use of this sum. You may invest it in any projects you like, and the profits you earn from them will be yours. You have only to ensure that you invest it soundly, so that the capital sum remains intact. But that is something I believe venture capitalists like yourself are very good at.

BANKER (*flattered, but still puzzled*): Aha! But what is in it for *you*? For a full 20 years you will be depriving yourself of the use of your own money, and you will receive no payments whatever during that period.

MERCHANT ADVENTURER (*thinking about Einstein's equations*): Let us say that I intend to age more slowly than you.

This is of course a highly idealised conversation, which I have only written in order to inform! No real-life banker would need to ask all these questions – except perhaps the last; he would already know the answers to them.*

There are nevertheless two important questions that seem to demand answers. How does the adventurer finance his *first* voyage, before he has earned any of his interest? One answer is that he will simply borrow it, using some property of his as collateral. He is in the position of one of those characters in P.G. Wodehouse stories who is short of cash but has a very rich uncle who is devoted to him and who will soon die. He knows for certain that he, or more probably his heirs who have remained on Earth, will be able to collect the interest and the capital at the end of the agreed period. A more sophisticated answer is that his principal banker (the person in the above dialogue) might lend him the money to undertake the voyage on the same terms as the second transaction. He might make it a condition

---

* The term 'banker' may require more precise definition. There are many different types of banks, from high street, or retail, banks to merchant banks. It is doubtful if any of these would be interested in funding starships. The kind of banker we are looking for is a *venture capitalist*, one who is prepared to fund long-term enterprises that he considers viable and who is prepared to wait years, even decades, to see a return on his investment.

that *he* is one of those to be lent money on the star traveller's return. It is true, of course, that the star traveller might be tempted to abscond in the depths of space, but with all that money waiting for him on his return, he would have no motive to do so. From the banker's point of view, it is a sound long-term speculation with immediate rewards from the use of the money he has been lent.

A second objection might be a gut feeling that this all seems too easy. Several people are earning huge sums of money either for sitting and waiting or else for travelling expensively. Surely someone is being swindled?

But this is not the case. Wealth is being created *by time*. No one is being swindled because no one is losing anything. The key sentence in my imaginary dialogue is: '*For a full 20 years you will have the use of this sum.*' This capital sum, while it is in the banker's hands, will stimulate the economy even if the economy is in recession. Moreover it will create countless jobs by starting innumerable profitable enterprises that will also enrich the banker. Indeed, one reason why such a mutually lucrative system cannot be started today is that no one has yet learned how to travel at close to the speed of light.* Time passes at the same pace for everyone. Nobody today wants to tie up his wealth for 20 years, because if he does so, he must wait 20 of *his* years before he can collect the interest. High speeds, however, change everything. Einstein's equations offer us a new financial universe.[8]

Suppose that one wished to finance a voyage to a planet circling the star Epsilon Eridani, 10.7 light years from Earth. In its barest simplicity, Warren Salomon summarises the workings of this scheme:

---

* Another reason is the existence of capital gains tax, which oppressively obstructs economic growth without producing any discernible gain. Although this tax typically brings in no more than a fraction of 1 per cent of typical national tax revenues, it is in many countries inexplicably kept alive. It is true that someone earning a trillion dollars from a capital gain would have to pay 'only' 20 per cent (the top rate in the United States in mid-1999) leaving him a capital profit of $800 billion. But he would certainly not relish having to pay the rest, $200 billion, to the Government! However there is hope that this tax will be abolished within the next few decades, mainly because an ever-growing number of voters own shares and naturally find the tax intolerable. See Paul A. Gigot, 'This Isn't What Marx Meant by Das Capital', *The Wall Street Journal Europe*, 22 March 1999.

> A rich man could put part of his portfolio at interest on Earth, invest the rest in an exploration company, and then climb aboard ship. After twenty-four years have elapsed on Earth, he returns only three years older, finds a potful of money waiting for him in the bank (his left-behind deposit having multiplied five or ten times), and he also owns the beginnings of a thriving business on Epsilon Eridani. After another trip or two, he's incredibly rich, still relatively young, and now his investment on Epsilon Eridani should be starting to pay off.
>
> Future trips can be made painlessly affordable by compound interest piling up during (subjectively) swift voyages. Commerce will be financed with interstellar letters of credit issued by branches of fabulously wealthy interstellar banks. Products will be paid for years in advance by departing star travellers making small deposits, which will swiftly inflate (via compound interest) over the time-dilated decades. Interstellar goods will be shipped to Earth and bought by wealthy, returning Rip van Winkles with a need for affordable, off-world obsolete goods. And, after a few centuries, when it all comes to pass, the star-travelling founders may still be alive to enjoy their much deserved dividends.[9]

It may be that what nature takes away with one hand she gives back with the other. It might at first sight seem a tragedy that faster-than-light travel, with all its supposed advantages, appears to be so rigorously forbidden. But is this really such a tragedy? What if we had the privileges of the fictitious crews of *Star Trek*, and could travel at up to 'Warp Nine', 1,516 times the speed of light? What financial advantage do they gain from it? None. Since the series ignores relativity, their time does not change, no matter how fast they travel. They are strictly tied to Newton's 'equable flow of time, without relation to anything external'. It is hard to see how these travellers could ever make a profit or how an infrastructure could be developed that would finance their voyages.

But time dilation will change some of the rules of economics. When combined with the wealth generated by the colonies themselves, it will bring into existence a new class of merchants who can afford to tie up their money for long periods, letting it remain under other people's control, who can in turn invest howsoever they please, retaining for themselves all profits from those investments. In short, it guarantees the creation of wealth. It is astonishing that so many space experts have failed to see its possibilities.

Chapter 9

A Fuel Like Magic

> There is no way in the universe to get more bang for your buck than to take a particle and annihilate it with its antiparticle to produce pure radiation energy. It is the ultimate rocket-propulsion technology, and it will surely be used if we ever carry rockets to their logical extremes.
>
> Lawrence M. Krauss, *The Physics of Star Trek*

Now for the practical problems of how to build starships and make them fly at relativistic speeds. The solutions are not always obvious.

There is an apocryphal story about a man who sets out in his car for a long journey through an empty wilderness. Since he must travel for days without any possibility of refuelling, he must take all his petrol with him. He stores 10 tons of it in the car, so that the fuel is vastly heavier than the vehicle that is to carry it. He starts the engine – but not surprisingly, the car is so heavy that it refuses to move.

The moral of the story concerns the design of starship engines. Their fuel must be efficient enough so that very small quantities of it must be able to do an enormous amount of work.

What kind of fuel? The resources of chemistry and physics are so great that there might, at first sight, appear to be many alternatives. Indeed, to someone of a poetical turn of mind, there might be a limitless number of ways to travel through space. Cyrano de Bergerac, the hero of Edmond Rostand's play of that title, boasts about his improbable and ingenious ways of travelling to the Moon:[1]

You wish to know by what mysterious means
I reached the Moon? Well, confidentially—
It was a new invention of my own.

I imitated no one. I myself
Discovered not one scheme merely, but six—
Six ways to violate the virgin sky!

As for instance – having stripped myself
Bare as wax candle, adorn my form
With crystal vials filled with morning dew,
And so be drawn aloft, as the Sun rises
Drinking the mist of dawn!

Or, sealing up the air in a cedar chest,
Rarefy it by means of mirrors, placed
In a icosahedron.

Again,
I might construct a rocket, in the form
Of a huge locust, driven by impulses
Of villainous saltpetre from the rear,
Upwards by leaps and bounds.

Three,
Smoke, having a natural tendency to rise,
Blow in a globe to raise me.

Four!
Or since Diana, as old fables tell,
Draws forth to fill her crescent horn, the marrow
Of bulls and goats – to anoint myself therewith.

Five!
Finally – seated on an iron plate,
To hurl a magnet in the air – the iron
Follows – I catch the magnet – throw again—
And so proceed indefinitely.

The ocean!
What hour its rising tide seeks the full moon,

I laid me on the strand, fresh from the spray,
My head fronting the moonbeams, since the hair
Retains moisture – and so I slowly rose
As upon angel's wings, effortlessly.

But to return to the more demanding conditions of the real world. We cannot travel to the stars by means of Cyrano's iron plate and magnet. Indeed, where speeds close to that of light are desired, all currently known methods of space travel turn out to be inferior. Today's spaceships are powered by chemical rockets, using a fuel of liquid hydrogen and liquid oxygen. But this will not do for journeys to the stars! For all the sound and fury they emit during the blast-off of a space shuttle (and although far more efficient than the kerosene that powers jet aircraft and the petrol that drives cars), they would be absolutely useless even for fast journeys between our own planets. For the latter, some form of nuclear propulsion will be needed. Various forms of atomic fission – the process that drives nuclear power stations – are being studied. They would indeed be about five million times more efficient than chemical rockets, making speeds for large spacecraft of tens of thousands of kilometres per hour.

However this also would be a snail's pace compared with the velocities needed for interstellar travel. Craft powered by thermonuclear *fusion* – bearing in mind that we have not succeeded in developing such power – would be much more efficient, 30 million times more so than the powerful kinds of fission energy.[2]

But even this is not powerful enough. The most energetic fuel in the known universe, and the ideal fuel for a starship engine, is that mysterious substance known as antimatter.

The dream of antimatter was latent in Einstein's special theory of relativity of 1905, although it was not for another quarter of a century that this was realised. His statement that $E=mc^2$ is perhaps the single most important equation that describes the nature of the universe. It shows that matter and energy are interchangeable. They could be compared to two currencies, like pounds and dollars, each of which can be converted into the other and back again.

Consider the literal meaning of the equation. E stands for energy, m for mass and c^2 for the *square* of the speed of light. The speed of light is 300,000 kilometres per second. Take that as a number, not a measurement, and call it 300,000. Its square, that number multiplied by itself, would be 90 billion. Thus it will be seen that a tiny amount of mass can produce a truly gigantic amount of energy. There is enough power locked up in a kilogram of matter to send 9,000 giant rockets to the Moon, or release the energy of 900 10-megaton hydrogen bombs. Even the mass in the smallest of our coins contains more energy than four atomic bombs.*[3]

Only antimatter can release such energies. Its possible existence was not immediately apparent after Einstein published his equation. It was only in 1930 that the British physicist Paul Dirac proposed that if the equation was correct, and energy could be converted back into matter, then *a new kind of matter* could be envisaged, whose atoms and electrons would have the opposite electric charge to ordinary matter. Matter and antimatter would be like mirror images of each other, resembling the two aspects of the sitting room in *Alice Through the Looking Glass* in which everything was identical but the opposite way round. If the two were ever to come into contact, extraordinarily violent consequences would ensue. They would merge, being of opposite charge, since atoms of opposite charge attract one another. But then, instantly, their combined masses would be converted into energy. To put this another way, imagine a man made of matter and a woman made of antimatter exchanging a kiss in a city centre. As their lips touch, the entire city is transformed into a smoking crater.

The reaction to Dirac's proposal in 1930 was that it sounded more like science fiction than a description of the real world.[4] But it was confirmed two years later by Carl Anderson, in his laboratory at the California Institute of Technology, when he photographed the track of an extraordinary new particle. It appeared to be identical to an electron, the particle that orbits the nucleus of an atom, except that when its motion curved under the influence of a magnetic field, it curved *in the opposite direction.*[5] In short, it was an anti-electron which, unlike an ordinary

---

* Many laymen greeted Einstein's equation with incredulity when he first produced it in 1905. 'What?' said one sceptic. 'Are you really saying that a fragment of matter smaller than a lump of coal can produce more energy than ten thousand regiments of cavalry? If this is true, why is it not apparent?' Einstein replied: 'Imagine a man who was fabulously rich. But if he never spent any money, no one would ever know how rich he was.'

electron, had a positive electric charge, a particle that was henceforth known as a 'positron'.

What is true of atoms and anti-atoms must be true of all large objects made of them, whether planets, stars or galaxies. If a star made of antimatter collided with another star made of matter the result would be like that of the kissing couple in the city, but on a vastly greater scale. The explosion could obliterate life on a planet thousands of light years away. *All* the ordinary matter and *all* the antimatter would be turned into energy in accordance with Einstein's equation.

Even in everyday life, we see the principle of mass energy conversion at work. Petrol drives a car along a road by turning the mass of its fluid into energy. The energy which the petrol creates cannot of course be converted back into more petrol – how convenient life would be if it could! – but this is only because petrol is an extremely inefficient converter. Less than a millionth of 1 per cent of it ever does any useful work. But when we are dealing not with a simple, rather volatile chemical like petrol, but with the unleashed powers of the universe, stupendous energies are revealed.

Some dim measure of Einstein's equation may be seen at work in the blast of a single nuclear bomb. But awesome and terrible as the effects may be, the equation is working with great inefficiency. Less than 1 per cent of the fuel is converted into energy, whereas a matter–antimatter explosion can be said to be working with an efficiency of *200* per cent. Why 200? Because *all* the matter and *all* the antimatter are *both* converted into energy.[6]

Fortunately for us, the universe is not composed in equal parts of matter and antimatter. Otherwise, the cosmos would be a ceaselessly exploding entity, and we would probably never have come into existence. But the universe was not always in this harmonious condition. When it was created by the Big Bang some 15 billion years ago, it probably consisted of half matter and half antimatter. Then, in the first few seconds after the creation, these two massed forces annihilated each other in a blast of radiation, and what remained of the rapidly expanding cosmos consisted mostly of ordinary matter.

But not all. Twice in the past decade, astronomers have observed huge clouds of what is undoubtedly antimatter spewing out from the heart of our Milky Way galaxy.[7] The ultimate cause is the furious turbulence at the galactic core, where billions of stars are crowded together, being periodically devoured by a giant black hole about five million times more massive than the Sun.* It is cosmic violence that creates antimatter from matter, just as more violence is unleashed when the two come into contact. As will be seen in a moment, antimatter is created when ordinary matter smashes into other matter at close to the speed of light.

Because of its tremendous explosive power, antimatter is thus the ideal fuel for a starship. This may sound like a dangerous procedure for the crew. But the ship would not blow up. Instead, it would be propelled by a series of comparatively small, controlled explosions, at a rate of several thousand detonations per second, aimed in the opposite direction to that of the vehicle's motion, accelerating it in the same way as a stream of gas pouring out of the exhaust of a space shuttle or a jet aircraft. It would similarly exploit Isaac Newton's Third Law of motion that 'for every action there is an equal and opposite reaction'.

With 100 milligrams of antimatter fuel – no larger than 10 grains of salt – a spacecraft could reach the Moon within hours rather than days, Mars within a week, or Pluto, the furthest planet in the solar system, in little more than a month.† It might be useful to make a comparison of the efficiency of various rocket fuels. A more technically advanced society than our present one will regard the chemical fuel used today to propel spacecraft as extremely inefficient. It takes enormous quantities of it to produce any appreciable energy. When men flew to the Moon, for example, it took a three-stage Saturn V rocket that was almost all fuel. The first stage weighed more than 1,800 tons and was nearly 100 per cent fuel. So also was the second stage, which weighed nearly 400 tons. The final stage, containing the command, service and lunar landing vehicles, weighed only 35 tons, the same weight as the

---

* The black hole at the centre of the Milky Way is currently believed to be 'asleep' because it has temporarily run out of stars to devour. The black holes that are believed to exist in the centres of all – or most – galaxies apparently go into these quiescent states from time to time.

† Another advantage of using spaceships with antiproton rocket engines is that they would be able to execute very sharp turns like a fighter aircraft. Craft with conventional fuels can execute only gradual turns because the required 'mission characteristic velocity' – the sum total of all the changes in speed needed for such a mission – would be much too great for the quantity of fuel they could economically carry. But whether the human body would be able to endure such very sharp turns is another question.

amount of fuel consumed by the first stage during its first nine seconds of engine burn, before it even left the launch pad![8] Here is a comparison of various fuels – the last of which of course has not yet been harnessed – in their ability to produce joules per kilogram (a joule is roughly the energy a person needs to climb one step of a staircase):[9]

Fuel	*Joules per kilogram*
Kerosene (in jet aircraft)	9.1 million
Chemical rockets	18 million
Nuclear fission	82 million
Matter/antimatter	90 *billion*

In terms of *today's* technology and resources, building an antimatter-powered starship would be an exceedingly costly project. About a dozen anti-hydrogen atoms have so far been created in atom-smashing machines like those in CERN (the Centre for Nuclear Research) in Geneva and Fermilab in Batavia, Illinois, and this has led to the half-humorous calculation – since price is based on scarcity – that the current price of antimatter is $300 billion per milligram.[10] Since, as will be seen, some 70 *tons* of antimatter will be needed for a one-way manned voyage to Proxima Centauri, the nearest star beyond the Sun, this puts the minimum cost of such a voyage, if it were made today, at about $20 billion trillion. An expensive trip indeed, since this is approximately a billion times more than the entire predicted wealth of the world in the year 2000.[11]

The scientist Robert L. Forward, one of those who dream of building antimatter-powered rockets, calls antimatter an almost 'magical' fuel because of its 200 per cent efficiency at unleashing energy.[12] But its price, at present, seems just as magical as its properties. To make interstellar travel a less exorbitant project, it will be necessary to reduce the price of a milligram of antimatter from $300 billion to about $10 million, a cost reduction of 99.997 per cent. As Forward puts it:

> This may still sound like a lot, but at ten million dollars per milligram, antimatter is already cost-effective for space propulsion and power. At the present subsidised price of a space shuttle launch, it costs about five million dollars to put a ton of anything into low Earth orbit. Since a milligram of

> antimatter produces the same amount of energy as twenty tons of the most energetic chemical fuel available, then a milligram of antimatter costing ten million dollars would be a more cost effective fuel in space than twenty tons of chemical fuel costing ten times as much.[13]

And so how will antimatter be produced in sufficient abundance to reduce the $300 billion price tag for one milligram of it to $10 million? The obvious answer is to produce it in large quantities, since abundance reduces price. But since antimatter explodes on impact with matter, we cannot go out and look for it; it cannot be mined like gold or copper. It has to be manufactured, and in carefully controlled conditions.

Antimatter is created when ordinary matter is made to collide – or collides naturally – with other ordinary matter at speeds close to that of light. Kinetic energy, the force unleashed by collisions, produces the equivalent of an explosion. As any motorist who has been in an accident knows, the greater the speed and mass of the impacting object, the more violent is the energy released on impact.*

Some numbers here are unavoidable. Nuclear physicists measure energy in electron volts. A television set produces about 20,000 electron volts. This means that when the electrons in the television tube are accelerated, they produce 20,000 electron volts of energy when they strike the back of the screen, enough to produce a clear picture on the front of it. But this is a very small voltage. The total amount of energy bound up in a proton, the largest particle in the nucleus of a hydrogen atom, is just under one billion electron volts. Thus, a proton that has been accelerated to a speed so great that its kinetic energy is *more than* a billion electron volts has more energy in its motion than in its internal mass.[14] The extra energy that it produces on impact will be changed back into matter. (Remember that energy and mass are interchangeable.) And what it changes back into can be antimatter.

---

* The kinetic energy equation, worked out by Lord Kelvin in the nineteenth century, is similar in form to Einstein's. It states that the energy released by the impact equals the mass of the impacting object multiplied by half the square of its speed.

At CERN and Fermilab, high-speed 'bullet' protons are slammed into a target made of thick tungsten or copper wire. The energy is released as a spray of gamma rays and atomic particles, and sometimes of *anti*particles as well. For unfortunately, the process is very inefficient. At Fermilab, only five antiprotons are produced for each 100 bullet protons, and in CERN even fewer are made. At this rate, producing even a milligram of antimatter will take 200 years.[15]

This inefficiency does not of course greatly matter when one is creating antimatter for scientific study rather than, as we are interested in doing here, in industrial quantities for fuel. The current process is also very expensive because it must be done in underground vacuum tunnels so that air does not limit the speed of the collisions. Moreover, the antimatter production rate is severely limited by the size of the tunnels and by the fact that the laboratories are in constant demand for other experiments. But antimatter production does not have to be carried out on a planet where there is air. It need not be carried out on Earth at all. As one scientist put it, 'there does not seem to be any fundamental reason why plants outputting many tons of antimatter per year might not be built, although it would not be practicable or desirable to locate such plants on Earth'.[16] For all the above difficulties would vanish if an antimatter 'factory' could be established in a place where there is a *natural* vacuum.

One's first thought might be of a huge man-made satellite, much closer to the Sun than the Earth's orbit, so that solar heat could be used as an energy source. But such a factory satellite, operating in conditions of extreme heat, would be expensive to maintain and to gain access to, especially since calculations have shown that, to produce antimatter in useful quantities, it would have to be at least 300 kilometres in diameter.[17]

The Moon then springs to mind. There is plenty of space on the Moon, there is no problem in getting access to it, and it has a nearly perfect vacuum. But the Moon's surface may not be hot enough to provide a powerful energy source for the ancillary industries that an antimatter factory will need. It is as far away from the Sun as we are. Its average daytime temperature is only 107 degrees. Yet there is another planet similar in many respects to the Moon and much closer to the Sun, where the daytime surface temperature is more than four times hotter – the world of Mercury. As Robert L. Forward puts it:

> Ultimately, we shall need an antimatter factory out in space or perhaps on another planet. This is not only because antimatter is very dangerous stuff to handle; the Earth's atmosphere also makes handling it very difficult. The ideal world for it would be Mercury, the hottest place in the solar system.[18]

When the Sun rises on Mercury, it gets hot enough to melt lead.[19] It increases to 450 degrees. Isaac Asimov once predicted that this would give Mercury the cheapest and most efficient source of energy in the solar system.[20] A solar power station on the surface of Mercury would be a huge energy converter, turning heat into electricity, light, and the machinery needed to set up and maintain factories. Mercury is not in fact the hottest planet in the solar system. That honour belongs to Venus. Venus, however, has an atmosphere 100 times thicker than Earth's, which would make it useless for antimatter production! But on Mercury there is almost a perfect vacuum.* It is also likely to remain an extremely lonely place in coming centuries, unlike the Moon, which will be the site of many industries. Mercury, because it superficially resembles the Moon and takes much longer to reach, is likely to attract only the most determined. As one science writer says of this bizarre planet, with its three-kilometre-high cliffs which stretch for hundreds of kilometres:

> Standing on the rim of a dust-covered crater, against the backdrop of an ink black sky, you shield your eyes from the deadly glare of the Sun, no longer a warm friendly feature of summer skies back on Earth, but a swollen globe three times larger. In the intense light, unfiltered by an atmosphere, you gaze at the lifeless planet.[21]

Mercury has other extraordinary characteristics that the Moon

---

* Mercury has about a trillionth of the atmospheric density of Earth. This is equivalent to saying that if all the air to be found in a terrestrial theatre auditorium were to be scattered evenly around Mercury, an area of 75 million square kilometres (compared with Earth's 495 million), it would account for all the air on the planet. Mercury's 'air' consists of minute traces of hydrogen, helium, oxygen, sodium, and potassium, the latter two probably being ejected from the crust by meteorite impacts.

lacks. Since parts of it face away from the Sun for very long periods, temperatures on those parts of its surface drop down to minus 173 degrees. This means a 600-degree change from the sunlit regions to the cold of the night-side that may make it possible to use temperature differentiation – in addition to the obvious need for nuclear power – as an ancillary source of energy. The presence of huge amounts of iron in Mercury's interior may also prove useful. And the planet need not be inhospitable to workers, despite dogmatic statements to be found in astronomy textbooks that 'human beings will never walk on Mercury'. Radar observations in 1991 showed that its north pole, being permanently shielded from the Sun, is actually made of ice – a remarkable discovery for the second hottest known planet in the universe! The presence of such huge quantities of water, making it unnecessary to take it there in spaceships, will greatly reduce the cost of setting up and running an industrial facility.

It can be a mistake to try to second-guess the future in too much detail, and it may be that there will be easier ways to manufacture antimatter in large quantities than setting up a huge factory on blazing-hot Mercury. We need to learn a great deal more about Mercury before being sure of the matter. The *Mariner*

10 spacecraft, which made four passes of Mercury in 1975 (missing the north polar ice) only surveyed 45 per cent of the planet's surface. It is also very difficult to observe Mercury through telescopes from Earth. Because it is so close to the Sun, it appears very close to the Sun in our sky. The only time to see it, a most difficult feat, is just after sunset or just before dawn.* And NASA's plans to send another mission may encounter political difficulties, since Mercury, so resembling the familiar Moon, as well as being so much more difficult and expensive to reach, lacks 'glamour'.[22]

Yet it is easy to visualise the operation in principle. Without the need for vacuum tunnels, since Mercury itself provides an almost perfect vacuum, matter could be made to collide, guided by man-made magnetic fields. The next important question is the kind of ships that the resulting antimatter will propel.

---

* In 1698, the astronomer Edmund Halley was on a voyage to observe Mercury crossing the Sun (the 'transit of Mercury'). He put in at Recife, Brazil, where he fell foul of the English consul, a Mr Hardwicke. This official declared that Halley's story of his intentions was too ridiculous to be believed and had him arrested as a suspected pirate.

Chapter 10

The Appearance of a Starship

> Everything beautiful and noble is the result of reason and calculation.
>
> Charles Baudelaire

> Hast thou given the horse strength? Hast thou clothed his neck with thunder?
>
> Job, 39:19

To turn the great French poet's statement around, one could say that a construction is beautiful and noble *because* it is the result of reason and calculation. If it is purely functional, and carries no adornment or appendages other than what it needs to work, then its sheer starkness will give it a certain beauty. The Futurist Society proclaimed this truth in 1909 when it expressed its admiration for one of the new-fangled racing cars, vehicles which less astute critics considered ugly and vulgar:

> The world's magnificence has been enriched by a new beauty: the beauty of speed. A racing car whose hood is adorned with great pipes like serpents of explosive breath – a roaring car that seems to ride on grapeshot that is more beautiful than the Victory of Samothrace.[1]

Manned starships are likely to be vessels with something of this quality. Their sheer functionalism will give them an extraordinary appearance, quite unlike the glorified fighter aircraft that we have learned to expect of starships from science fiction.

Most people who think of starships imagine a vessel like the *Enterprise* in *Star Trek*, a ship whose basic design is not unlike that of a jumbo jet. There is a huge pancake-shaped disc to house the crew compartments and flight deck, and below this, and sensibly separated from it, are two enormous rocket nozzles pointing to the rear.

The basic idea of the *Enterprise* is quite straightforward – apart from the mysterious lack of any structures big enough to be the fuel tank and engines. Its antimatter engines (which we never see but only imagine) work in principle like jet engines. But instead of expelling gas as the jets of a kerosene-powered aircraft do, they create continuous massive explosions of matter and antimatter at the rear of the ship, driving it forward. Both the jet engine and the *Enterprise* are thus obeying Isaac Newton's Third Law that 'for every action there is an equal and opposite reaction'.* The only difference is that the former does it by expelling gas, and the latter by creating explosions.

So what can possibly be wrong with this? To make a real-life starship fly, we only have to add a radiation shield to protect the crew from stray gamma rays, an engine, coolers to prevent the ship from becoming molten, and a device for confining the antimatter fuel so that it does not react with matter prematurely and blow up the ship, and it would work.

Wouldn't it?

Well it would, from an engineering point of view, but not from the point of view of economics. The difficulty is that holding the rocket nozzles, engines, radiation shield and the rest in place would require a structure of massive and enormous girders. The radiation shield in itself would weigh many thousands of tons. All this would add intolerably to the mass of the ship and the amount of fuel needed to accelerate it.

Such a design would surely make interstellar travel prohibitively expensive. According to one scientific study which could be

---

* The acceleration of the *Enterprise* is not always convincing, to put it politely. In one episode the ship was orbiting a planet at 28,000 k.p.h., when the captain received an urgent message to the effect that he should be somewhere else. In a blurring sequence, the ship immediately accelerated to 'Warp Five', which is 125 times the speed of light! Even if this was scientifically possible, the vibration caused by such unimaginable acceleration would shake the ship to pieces and kill everyone on board.

called accurate but unimaginative, a manned vessel shaped like the *Enterprise* (but without its 'warp' capabilities) would need 400 million tons of antimatter fuel to make a journey to Alpha Centauri.[2] Even at Robert Forward's predicted antimatter price of $10 million per milligram, this would entail a voyage costing $4 trillion trillion, which is 200 billion times the expected gross world product in the year 2000!

Surely it is possible to travel more economically than this.

Indeed it is. These problems largely vanish if instead of having an engine behind that *pushes* the ship, we have one in front that *pulls* it.

There is a famous photograph of the American rocket pioneer Robert Goddard, taken in 1926, standing in his Aunt Effie's cabbage patch at Auburn, Massachusetts, beside his experimental liquid-fuelled rocket – the first ever to be launched – which was to fly to a height of 12 metres.[3] It was a very primitive contraption, put together in a garage. It stood barely taller than himself and consisted simply of a 'launch tower' made of some tubing, the rocket engine above, and *below*, fixed to it by two rods, a chamber with a nose cone (the 'payload'), intended to show that it could carry a cargo. Only in later versions which could fly much higher did Goddard add a parachute to ensure that the payload landed safely without being broken or destroyed. He typically assembled his rockets from any bit of hardware that would do the required job – a child's wristwatch, a length of piano wire or a car sparking plug.[4]

His 1926 rocket may have been unbelievably crude compared with a modern space rocket, but in one respect it was far more advanced. The engine *pulled* the payload rather than *pushed* it. Goddard's design has been long abandoned. Perhaps because the extra weight involved in a 'pushing' rocket did not greatly matter to people building a comparatively small Earth-to-orbit vehicle, engineers have long since put the engines in the rear – or at the bottom – in a craft lifting off vertically. (An aircraft jet engine, which is a form of rocket, could not possibly be anywhere but on the plane's tail or wings since otherwise, when taking off, the passengers might be bounced violently along the runway!) But now two scientists interested in starship design, Charles Pellegrino and Jim Powell, have re-invented Goddard's configuration. Rule Number One, they say, is to keep the mass of the ship as low as possible. 'Even an added gram means extra fuel.' They have proposed a ship whose engine would tow the cargo at the end of a tether about 10 kilometres long, as a motor boat tows a water skier.[5]

It will be seen in a moment why the tether has to be twice as long as an airport runway.

On the cover of Pellegrino's 1993 science fiction novel *Flying to Valhalla*, there are artists' impressions of these starships with the engines towing the passenger compartments. They look absolutely weird – long, thin beams of machinery streaking through the heavens like arrows. Being so gigantic, and depending on such a long tether for their propulsion, they will of course be built in space, using construction materials from an asteroid. Jovan Djordjevic illustrates this chapter with such a design (see opposite). Only an observer familiar with Baudelaire's precept would realise that he was looking at a man-made spacecraft. For there is almost nothing there that does not *need* to be there.*

There are no towers, no complicated and heavy structures, none of those immense struts and girders so familiar in engineering edifices whose sole purpose is to hold other structures in position. There are only the crew quarters, which appear as a large sphere. They do not have to be spherical, of course; they could just as easily, and perhaps more conveniently, be a rectangular block that would be capable of being expanded.

A word about these compartments. However vast and spacious they may be, they must weigh very little if Pellegrino and Powell's Rule Number One is to be observed. We can do much better than the titanium alloys used in jet aircraft hulls which are as strong as steel and half its weight.

In what is becoming the golden age of exotic new materials, substances will soon be in use that are as tough and as resilient as steel and far lighter than titanium alloys. The X-33, a new

---

* Nineteenth-century engineers would not have approved of the design of Pellegrino–Powell starships. Whether they were making sewing machines or suspension bridges, they had a passion for unnecessary decoration which, instead of adding dignity, added only weight and cost. They had no faith in the aesthetics of unadornment. They loved gold and cast-iron filigrees, columns, statues, and engravings. Isambard Kingdom Brunel had planned elaborate Egyptian-style decoration for the Clifton suspension bridge over the Avon in Gloucestershire which would have added significantly to its cost and expense. One doubts whether he would have liked its stark, undecorated beauty had he lived to see it.

spacecraft now being built by Lockheed Martin to replace the space shuttles, will use graphite composites and aluminium–lithium resins that are not only much stronger than steel but *five times* lighter.[6]

We can do even better than this if the crew compartments and much of the rest of the ship can be constructed by nanomachines and put together atom by atom. Nanomachines, I should explain, are future devices in which a great deal of hope has been invested. They will be, literally, machines the size of bacteria that will take atoms of different elements and assemble them into molecules, or back again into atoms and then into fresh molecules. In short, one could make almost anything out of the most unpromising materials so long as the right atoms were present. One cannot tell of what exotic new materials these parts of the ship will be made; only that by today's principles they will be unimaginable. They may even be thousands of times lighter than steel.[7]

But it would be too much to hope that these materials will be resistant to extreme radiation. They may be, but we must be as conservative and cautious as possible and assume the worst. And so the crew must be protected from the mortal dangers presented by their own engine. The ship will be accelerated for more than a hundred days to reach its coasting speed of 92 per cent of the speed of light. A phenomenal amount of energy will be expelled from the engine exhaust towards the rear of the ship. Having the engine in front of them – even 10 kilometres in front of them – there is an obvious danger of the crew being blasted by its discharges. They could be in peril of annihilation from streams of deadly gamma rays from the antimatter engine. As Pellegrino puts it, 'riding an antimatter rocket is like riding a giant death-ray bomb. An unshielded man standing a hundred kilometres away from the engine will receive a lethal dose of gamma radiation within milliseconds.'[8] He and Powell have come up with a simple but ingenious answer to this problem. A hundred metres behind the engine, to shield the crew from the gamma radiation, they propose putting a block of the metallic element tungsten.

Tungsten, named from a Swedish word meaning 'heavy stone', is an exceptionally strong metal with the highest melting point and the highest tensile strength of all metals. For this reason it is used in electric furnaces, in the rocket nozzles and atmospheric re-entry heat shields of spacecraft, and in other extreme high-temperature applications.[9]

It is also comparatively cheap, being as abundant in the Earth's crust as tin, and for the convenience of starship manufacturing economics it is plentiful on the gravity-weak Moon and probably also in the asteroids.[10] Most important, it can absorb gamma rays. It will reduce their energy by a tenth for every two centimetres of tungsten they pass through.[11] A block of tungsten 20 centimetres thick – ten times two centimetres – will therefore reduce the energy of the gamma radiation from the engine by a factor of 10 billion. And if this is not reckoned to make conditions safe enough for the crew, a block of tungsten just a little thicker, say 30 centimetres, will, by the same arithmetic, reduce gamma radiation by a factor of 1,000 trillion.

The tungsten shield, moreover, does not have to be at all *wide*. By placing it 100 times closer to the engine than to the crew compartments, its width need only be a hundredth of the 'shadow' – the source – of the gamma radiation, and the weight of the shielding will hence be trivial. So far, Pellegrino and Powell obey their Number One Rule.

What other catastrophic dangers might there be during this long, lonely voyage? The other most deadly one is the risk of collisions. Interstellar space is not absolutely empty, as many people have imagined, but is full of clouds of gas and dust.

These clouds are trillions of times more tenuous than terrestrial air, but are dense enough to be dangerous to starship travellers.[12] At a speed of 92 per cent of the speed of light, a collision even with the tiniest object, no larger than a grain of dust, would be like the impact of a grenade. There would be an explosion equivalent to that of 100 kilograms of TNT.[13] And a larger one, the size, say, of a hailstone, would have the impacting force of an exploding torpedo and could obliterate the ship.

The size of a *hailstone*? This may sound extraordinary. There is no phenomenon on Earth that prepares us for such a calamity. If we walk along a beach, and a sudden wind springs up, we do not worry about being injured by flying grains of sand. They may temporarily blind us if the wind is strong, but they are not going to punch holes in us. But this is only because the speed of the sand grains will be trivial, no more than 10 k.p.h. or so. But when the impact occurs at 92 per cent of the speed of light, the energy of the collision will be unimaginably violent.*

There are dangerous and unseen rocks and shoals in interstellar space just as there are in the oceans. Starship designers and interstellar navigators will have to take note of several disturbing astronomical observations in the past decade. As Eugene Mallove and Gregory Matloff point out in their book The *Starflight Handbook*, we are still far from having an accurate knowledge of conditions in the interstellar medium, even that tiny part of space that lies between the Earth and Alpha Centauri.

---

* In the equation of kinetic energy worked out in the last century by Lord Kelvin and James Clerk Maxwell, the energy of the collision will equal the mass of the grain multiplied by half the square of the speed of the ship.

We see this by observing quasars, brilliant, star-like objects at the edge of the universe. They are 'blinking', as if something unseen is passing in front of them. (One can see a similar effect, on a local scale, by looking at the rings of Saturn as it passes in front of a distant star. The light of the star blinks on and off as Saturn's rings obstruct it.) The implication is that these 'somethings' may be failed stars, too faint to be seen from Earth, surrounded by clouds of dust even thicker than the ones we have been discussing.

The danger will come not from the faint stars, which will be easily seen by the ship's radar and avoided, but from the dust itself. It may even be necessary to chart courses around these perilous reefs of the cosmos if starships are to travel close to the speed of light.[14] One cannot successfully design a ship without also studying the surroundings in which it will travel. Indeed, as Mallove and Matloff put it:

> Starships will have a significant interaction with the tenuous broth of molecules and other particles between the stars. The higher the speed, the worse the problem. To design a starship without thorough knowledge of the interstellar medium would be almost as foolish as trying to build an aircraft without considering the properties of Earth's atmosphere, or planning an ocean-going vessel without thinking about the properties of sea water. One should not pretend, as some have, that vacuum alone reigns supreme in interstellar space.[15]

What can be done in general about the perils of dust and gas? The ship could of course travel more slowly, which would reduce the danger. The captain of the *Titanic* was confronted by a similar dilemma. He accelerated to full speed in order to arrive in New York a day early and win the Blue Riband. But here we cannot slow down as that captain ought to have done. For to do so would increase the voyage time not by a day, but by many decades, which would defeat its own object. Instead, some sort of anti-collision shield must be erected in front of the ship. A 'passive' shield, some sort of protective screen, would be useless, however strong it was, since it would be destroyed itself by impacts.[16]

An 'active' shield must be designed, a device that ejects matter ahead of the ship and destroys or scatters obstacles in its path.

The most promising way to do this is to exploit the vehicle's excess heat. There will be plenty of excess heat to exploit. A matter–antimatter engine will produce enormous amounts of it. Much of this will accumulate in the tungsten block, and a great deal in the reactor and in the region of the exhaust nozzles.*

This excess heat must be ejected in front of the ship in the form of fluid, in streams of hot droplets that will ionise all atoms that they encounter, stripping away their electrons from their protons. This will work because at temperatures higher than about 10,000 degrees, *all* matter is turned into plasma, a fourth state of matter, matter without atoms, beyond the three states of matter we are familiar with, solid, liquid and gas.[17] Plasma will present no obstacles. As Pellegrino puts it, 'the rocket itself then shunts the resulting shower of charged particles – protons and electrons – off to either side, much the same as when a boat's prow pushes aside water'.[18]

There is a nagging final thought when one considers this Leviathan of a ship, so massive and so many kilometres long. Its dimensions make our biggest aircraft seem like cockle shells in comparison. Where will it be built, from where will it be launched? It is obviously absurd to suppose that it could be constructed on Earth and then launched into space. It is even far-fetched to imagine that it could be safely assembled in Earth orbit or any such crowded region. It certainly could not be flight-tested there. A single blast of its main engine would unleash more energy than mankind has released in the past 400 years.

It seems probable that the first manned starship will be constructed comparatively far out in the solar system, high above the dangerous asteroid belt, somewhere beyond the orbit of Jupiter, where the Sun's radiation is negligible, and the outside temperature virtually the same as in interstellar space. A period of several years should suffice to ensure that all systems are working reliably.[19]

★ ★ ★

---

* Indeed, in the core of the reactor, this heat may be so great that unintended nuclear fusion reactions may take place; the crew may even find themselves with an unexpected bonanza of gold, a useful medium for currency on their destination planet. This is perhaps speculative, since gold is created at several billion degrees in the cores of giant stars, but I throw out the idea for what it is worth.

Now for the final all-important problem. After more than two years of coasting (ship time), the Alpha Centauri system will be only light months away, and it will be time for the ship to slow down from its tremendous cruising velocity. Failure to do so in time would be calamitous. There is likely to be so much debris – moons, dust clouds, and asteroids – in the destination star system that the ship would be certain to collide with something, a collision that even the fluid shield could not prevent. If it collided with this massive material at 92 per cent of the speed of light, the ship would never be heard of again.

In their proposed method of reducing speed to avert such a catastrophe, it is possible – just possible – that Pellegrino and Powell may be breaking their Rule Number One, to keep the ship's mass as low as possible. They suggest that the ship should have a *second* engine in its rear that stays inert during the coasting period of the voyage. When the time comes to decelerate, the ship would turn itself around and the second engine would fire, acting like a retro-rocket.[20] It would be like that moment when a jet aircraft lands on the tarmac at 300 k.p.h.; the engines fire in reverse thrust with a thunderous roar and the plane slows to taxiing speed. However, it would not be practical to expect an antimatter engine to fire in reverse. An enormous and costly amount of energy would be needed to swing it round so that it faced the opposite direction and fired forwards instead of backwards. And such an engine will be a highly dangerous mechanism that no one will want to get too close to, even when it has been switched off! Clearly it would be much easier to swing the *ship* around. Hence the perceived need for a second engine in the rear.

It is just possible that there could be a much cheaper alternative to this idea. An enormous amount of mass could be saved if we could dispense with this second engine and find some other means of braking. A more practical alternative idea might be to erect a sail that will be caught by the stellar wind of the destination star, when it is close enough, and slow the ship down.[21]

I must explain this briefly. Johannes Kepler, in the early seventh century, made the observation – which puzzled many people at the time – that the tails of comets always point *away* from the Sun, irrespective of whether the comet is approaching the Sun or receding from it. He concluded that the Sun exerts an invisible force, a solar 'wind', on the comets, vaporising their ice and pushing it away from them in fiery streams. The Sun in fact emits two roughly constant winds consisting of streams of atomic particles, one travelling at some 800,000 k.p.h. and emanating from the solar equator, and another travelling twice as fast coming from the poles.[22] There is nothing gentle about these particle streams. They boil off the Sun's surface with such fury that their effect is felt far beyond the confines of the solar system. The region known as the 'heliopause', where their speeds slacken to a few hundred kilometres per hour, is about seven billion kilometres from the Sun, a billion kilometres beyond the orbit of our most distant planet.

Spaceships, therefore, do not necessarily need engines if they are travelling in the close vicinity of a star. They could erect a solar sail and be pushed by this wind. (See Chapter 11 for more details.) Within our own solar system, a small spacecraft, bearing a sail of one square kilometre in size and a hundredth of a millimetre in thickness, could accelerate itself away from the Sun to the comparatively slow speed of more than 110,000 k.p.h. within the space of a year.[23] Recall that Alpha Centauri A and Alpha Centauri B are stars highly similar to the Sun. It is therefore reasonable to expect that they each have stellar winds of similar strength. If this is so (a detail which astronomers will have to ascertain before the ship departs!), a ship *approaching* such a star system from a great distance could slow down by spreading its sail as soon as the wind from the star became significantly strong.

But there is a way of braking that is even more efficient than spreading a sail. A ship containing hundreds of people and all their equipment would need a braking sail of thousands of square kilometres as it approached the Alpha Centauri system. Even if the sail was only a fraction of a millimetre in thickness, a quarter of the thickness of the plastic in a domestic rubbish bag, it would still be very massive when put into storage and folded up. And an accident when deploying it, such as a tear in its fabric, could mean that the ship continued at full speed, with catastrophic results.

Robert Zubrin, in his book *The Case for Mars*, suggests that a *magnetic field* would serve much better than a sail. The people of Earth have long been familiar with the effects of magnetic fields, even if they have only recently fully understood them. They appear in the multicoloured spectacle and ominous crackling sounds of the Northern Lights, which terrified the Norse people of the Middle Ages, who saw in them the war god Thor and his flashing eyes and rattling chariot wheels.[24]

They are in fact a manifestation of the Earth's magnetic field – or 'magnetosphere' – which protects us from the solar wind. As Zubrin says: 'If the Earth's magnetosphere blocks the solar wind, it must be creating drag, and therefore feel force as a result. Why not create an artificial magnetosphere on a spacecraft and use the same effect for propulsion?'[25] The ship, when the time came to brake, would simply deploy a rugged superconducting cable that would form a stiff loop surrounding or adjacent to the ship in which it would generate a large magnetic field.

This field would have a hundred times the thrust-to-weight ratio of a sail, exerting a force of a million amps of current on each square centimetre of the field.[26] The other great advantage of a magnetic field over a sail is that the former could be made arbitrarily big, without the on-board space constraints of the furled sail. Nor would it need any electric power to keep it going during the hundred days or so that the ship would take to slow down. Because the superconducting cable would have no electrical resistance, no additional energy would be needed to maintain it.* And if, as one must expect, the flares of the Alpha Centauri stars are as dangerous as the Sun's can be, the magnetic field will of course shield the crew completely against them.[27]

When introducing the idea of using the radiation of the destination star's wind as a means of slowing down the ship from relativistic to interplanetary speeds, I was intentionally vague. Experts are divided about whether it will provide sufficient braking power. Pellegrino and Powell believe that it will not, while others assert that it will.[28] Nobody, as far as I know, has done any detailed calculations to resolve the matter. It seems that in expert discussions of starships, the problems of acceleration and cruising speed have so far proved much more intellectually interesting than the equally important problem of how to slow down.

At this point, it seems fruitless to write any more about the detailed appearance of a ship that may not be built for more than

---

* It is not possible to test this technology today, in space or on Earth, because the necessary superconducting cables do not yet exist. But such rapid progress is being made in superconducting technology that it can only be a matter of decades before they do. They will surely have many applications apart from space travel.

another century. There will be many relevant inventions and developments far beyond what we know or can imagine. Since it is impossible to know what limits there will be to the lightness and strength of new materials, it is impossible to know the mass of the ship, or the quantity of antimatter required to accelerate it to 92 per cent of the speed of light. But we can be sure that, because the engine will be pulling rather than pushing, the necessary fuel will be the merest fraction of the 400 million tons in the study I cited earlier.

But will the ship obey Baudelaire's precept and be spectacular by being purely functional? Perhaps. Consider this description:

> . . . its white tanks gleaming against a hazy grey sky, its vertical tracery of flues and tubes in delicate silhouette beyond. Forms as complex and perhaps as beautiful in their intensely twentieth century way as the cathedrals of earlier eras.[29]

But this is no fine work of art, no magnificent museum piece. It is an architect's description of the Rotterdam oil refinery.

Part Two

THE INFINITE JOURNEY

Chapter 11

Rocketless Rocketry – and Other Methods

Suppose the chariot of the Sun were given to you. What would you do with it?

Ovid,
Metamorphoses

Will you hoist sail, sir? Here lies your way.

William Shakespeare,
Twelfth Night

Rocket power need not be the only means of space travel. There are other proposed methods of propelling a starship that do not involve rockets at all. Robert Forward has coined the phrase 'rocketless rocketry' for such concepts. If they prove practical, the advantages would be tremendous. Without the mass of fuel or main engine, even an engine that pulls rather than pushes, the ship will be vastly lighter, and much less energy will be needed to accelerate it.

These alternative methods are much simpler than antimatter propulsion, and they require such (comparatively) modest technology that they could be put into operation almost within a century. But at the same time, as will be seen, they could prove to have serious drawbacks which are political and psychological rather than technical.

Their basic idea can be seen with a terrestrial analogy.

One can play a game at fashionable seaports called 'crossing the ocean'. The idea is to stroll along the quayside, glance at the medium-sized yachts, and decide which of them could cross the Atlantic without foundering.

The answer is sometimes surprising. The large and luxurious motor launches, or 'gin palaces', might never be capable of doing so. Their problem is not only their lack of a keel, but that they are too small to carry enough fuel. Sailing boats, on the other hand, even much smaller vessels than the gin palaces, would have no such problems because – except to get in and out of harbours – they do not need to carry any fuel.*

To take this analogy into space, it may be possible to *sail* to the stars.

Winds come out of the Sun which, unlike the winds that fan the oceans, are almost unvaryingly constant. As I explained in the last chapter, three powerful streams of radiation pour continuously from our parent star. One, from the solar equator, moves at nearly a million kilometres per hour. And two others, from the solar poles, travel at almost double this speed. Theoretically, a solar sail, or rather a magnetic field deployed by the starship – the very same magnetic field that the ship will later use to decelerate when it approaches its destination star – could catch this wind and be accelerated by it to a significant percentage of the speed of light.

But it will not be quite as simple as that. Calculations have shown that while the wind from the destination star may be strong enough for deceleration over a very long period, the wind from our own Sun is *not* strong enough for the required *acceleration*. Its strength decreases with distance. Beyond the orbit of Jupiter it becomes too feeble to provide sufficient power.[1] It seems that sailing with solar power will be practical for flights within the solar system, where much lower speeds will be used, but that this power will be vastly insufficient for the much higher velocities demanded by interstellar travel.

There is a joke in *Gulliver's Travels* about a man who spends eight years on a project to extract sunbeams from cucumbers to make portable heaters, an absurd quest indeed for something so

---

* I have seen single-masted sailing boats not longer than 10 metres anchored in Antarctic bays. To get there, they crossed thousands of kilometres over the roughest ocean in the world. It would be inadvisable to attempt such a voyage in a gin palace.

fleeting and insubstantial.[2] The situation becomes very different, however, if sunlight can be artificially concentrated. A controlled sunbeam can be a projectile of devastating power, as seen, for example, in the James Bond film *The Man with the Golden Gun*, in the scene in which the villain uses one to blow up an aircraft at a range of more than 100 metres. As John Mauldin argues in his book *Prospects for Interstellar Travel*: 'What if a beam of sunlight could be collected and directed at the starship over a longer time and distance? A large curved parabolic mirror can collect sunlight to form a tight beam which can travel a long distance before spreading too much.'[3]

The idea is that such a mirror, 100 kilometres wide, would be placed as close as possible to the Sun, where sunlight reaching it would be at its strongest, and then aimed at the receding ship's magnetic field.

One can imagine such a mirror being in close orbit round the Sun, but being an orbiting piece of machinery would make the task of building and maintaining it exceedingly complicated and risky.

A far better place would be on the surface of Mercury which, as noted earlier as a site for antimatter production, lies nearly three times closer to the Sun than does the Earth, and is the hottest accessible place in the solar system.* Safe on a planet's surface, there would be no risk of its being lost in space or suffering other disasters.[4] And at Mercury's north pole, a region without sunlight, there are substantial quantities of ice which would help to provide the industrial infrastructure needed to build and maintain the mirror.

It has been calculated that a mirror of this size, if polished as smoothly as possible, would generate a beam of light that would become no wider than 200 kilometres out *to a distance of one light year*.[5] This would provide more than enough power to accelerate a starship, with a suitably large magnetic field, to relativistic speeds.

Obviously, there will be problems concerned with aiming the beam so that it always reaches the starship and does not dissipate its tremendous energies wastefully into space. Bearing in mind

---

* Mercury orbits the Sun at an average distance of 58 million kilometres compared with Earth's 150 million.

Mercury's complicated rotation and orbit, this may present formidable difficulties, and ingenious contrivances will have to be made to solve them.

But – and here is the essential point – if occasional mistakes are made, and the beam temporarily loses the ship, it does not greatly matter! For the purpose of the beam is to give the ship continuous acceleration. If this acceleration is for some reason momentarily interrupted, no great disaster will ensue. The ship will continue to cruise at the speed at which it was travelling when it was last in contact with the beam. There is no medium in the vacuum to slow it down by friction. It will not behave like an aircraft or a boat that slows down when its accelerating mechanism has been switched off. And acceleration will once more resume when the beam again makes contact. It will be a system with plenty of margin for error.

There are other variations on the idea of the 'beamed' starship. The propelling beam does not have to consist of pure, concentrated sunlight. It could be an all-powerful laser. Or it could be a mighty stream of atomic particles, or 'pellets' as they are sometimes called, generated by a nuclear power station. Again, the source of them would be a giant installation on Mercury, because it is the most convenient place; and again, their energy would be derived from the Sun, whose power is for all practical purposes unlimited.[6]

The concept of a beamed starship, however, convenient and simple though it is, contains one potential major flaw. The voyage would not be under the crew's control. They would be dependent on the good will of other people back in the solar system whose philosophy of life might differ from their own. It is perhaps not too cynical to suggest that at some time when the beamed acceleration had begun, the crew could all too easily expect to receive a message saying something like this:

> Dear Sirs,
>
> It is with the greatest regret that we must inform you that due to [such-and-such an event] our operating costs have risen by 5 per cent. We have no choice but to pass on to you this lamentable and unforeseen extra expense. Therefore, unless you undertake immediately to compensate us, we shall have no alternative but to switch off the beam.

The unfortunate crew would be compelled either to abandon their expedition, or else submit to this blackmail, knowing that it might be continually repeated until they at last attained their cruising velocity and became independent of the greedy people on Mercury. Until then, they would be at their mercy.[7]

So much for rocketless rocketry, a brilliant concept if only it could be guaranteed not to run foul of the frailties of human nature. Fortunately there are other proposed methods of interstellar propulsion in which the crew would be in charge of their own destinies and could not be blackmailed.

None of them produces acceleration as powerful as the anti-matter engine, but they are much cheaper and for that reason worth examining.

The British Interplanetary Society produced in the 1970s an impressive scheme by which a spacecraft could be sent to Barnard's Star, six light years away in the constellation of Ophiuchus, cruising at 12 per cent of the speed of light. Named Project Daedalus after the great inventor and mythical pioneer of flight, it would be accelerated by the continuous explosions of miniaturised hydrogen bombs in its rear, each the size of a tennis ball, exploding at a rate of 250 detonations per second.[8]

The statistics of Project Daedalus sound indeed startling. Using as its fuel the isotope helium-three from the atmosphere of Jupiter or Saturn, its speed would be 60,000 times faster than Concorde's, and 3,000 times faster than the *Apollo* ships on their way to the Moon. But even at this great velocity, it would take such a craft nearly 40 years to reach Alpha Centauri. This may be suitable for preliminary, unmanned missions to a star, but for transporting people, something more rapid is needed.

Instead of carrying its fuel, which, as we have seen, adds vastly to the mass of the vehicle, why not take its fuel from space itself? In 1960 the engineer Robert Bussard proposed his interstellar ramjet, or 'Bussard ramjet', which would do just this. Interstellar space is filled with tenuous clouds of hydrogen gas, and this could make an ideal starship fuel. In a sense, such a ramjet-driven vessel would be a form of 'sailing ship', with apparently limitless range, since the source of power would be *outside* the vehicle. Just as an aircraft jet engine breathes in air from the surrounding atmosphere in order to expel it at the rear, so would Bussard's ramjet suck in hydrogen from tens of thousands of kilometres around the ship to feed into a thermonuclear fusion reactor and use as fuel.[9]

The ramjet would attract the hydrogen by magnetism. This may sound peculiar, since normally iron and nickel are the only chemical elements that can be attracted by a magnet. But all elements become magnetically attractive when their atoms have been stripped of their electrons and they have been *ionised*, turned into plasma, the fourth state of matter, beyond solid, liquid and gas. To do this, it is necessary only to heat them. Hydrogen becomes plasma when it has been heated to a little over 5,000 degrees. To heat up the interstellar hydrogen, a Bussard starship would spray it with laser beams. Then, the hydrogen having become attractive to a magnetic field, it would be gathered up by a 'magnetic scoop'.

The hydrogen that lies between the stars is admittedly very tenuous, with a density of about one hydrogen atom per cubic centimetre, and Bussard's magnetic scoop would have to range far afield to gather sufficient fuel for the ship's reactor. For this reason, it is envisioned that the scoop would have to be so powerful that it gathered ionised hydrogen from a region at least 300,000 kilometres wide on either side of the ship. This may sound a formidable proposition, but, as Bussard saw it, it was a great deal easier than *carrying* all that fuel on board!

Elegant and simple though Bussard's concept was, later calculations showed that it contained a serious flaw. For as the scheme stood, there would be a 'traffic jam' of incoming fuel. Only about 1 per cent of the hydrogen could actually be used as fuel. The rest would pile up in front of the vehicle, literally clogging up the intakes. The ship would expend so much energy pushing its way through this useless hydrogen that it would never be able to accelerate to a reasonable speed.[10]

In 1974 the rocket designer Alan Bond studied Bussard's scheme, considered its flaw, and created a version of his own. He decided that the intake of hydrogen from space should be greatly reduced, and be used only to keep constant a quantity of hydrogen fuel that the starship would already be carrying. There would be *two* rocket exhausts, one from the ship's fusion engine, and the other from the interstellar hydrogen being thrust out of the rear *by means of* the energy from the fusion reaction. His was thus a hybrid means of propulsion, employing two separate systems that worked simultaneously.[11]

Bond called his version the Ram-Augmented Interstellar Rocket, abbreviated as RAIR, and there is a consensus among engineers that, depending on the combination of elements being used, RAIR's performance could be up to three times better than that of Bussard's system.

Nuclear fusion presents many exotic possibilities that could improve the efficiency of a RAIR. Hydrogen was once the only chemical element in the universe. Now there are 94 natural elements, all created by fusion, when the atoms of one of them, heated to tremendous temperatures in the cores of giant stars, are converted into the atoms of another. A reaction involving the elements boron and lithium, it has been suggested, could increase the ship's cruising speed to up to 70 per cent of the speed of light.[12] This is considerably less efficient than the antimatter engine of Pellegrino and Powell, but it is also considerably cheaper.

What kind of starship propulsion system, then, will mankind eventually use? Antimatter engine, sail, RAIR, Daedalus hydrogen bombs, or what? Probably all of them at various stages, and perhaps two or more of them in conjunction. On Earth, every conceivable kind of ship design has been successfully used, from quinquireme to hydrofoil, from raft to submarine, and there will probably be parallels in the evolution of starship design. Designers will always be juggling and compromising between the three precepts of speed, science, and economy.

No method of propulsion, however, will be of the slightest use if the crew do not know where they are in space; in other words, if they cannot navigate.

Chapter 12

The Star on the Starboard Beam

'Look,' whispered Chuck, and George lifted his eyes to heaven. (There is always a first time for everything.) Overhead, without any fuss, the stars were going out.

Arthur C. Clarke, *The Nine Billion Names of God*

About, about, in reel and rout
The Death-fires danced at night;
The water, like a witch's oils,
Burnt green, and blue, and white.

Samuel Taylor Coleridge, 'The Rime of the Ancient Mariner'

The Ancient Mariner was in great navigational difficulties. Having shot the albatross which brought good fortune, he was cursed by evil luck. The sea began to blaze with unaccustomed colours so that he found it impossible to tell where he was or where he was going.

It will be something like this for the crew of the first starship that ventures out of the solar system. They will encounter conditions that, to the inexperienced eye, will appear so outlandish and bizarre that there is every danger of becoming hopelessly lost, of not having the faintest idea in which direction to proceed, and of running inexorably out of fuel, food, and air.

Once a navigational error has been made, it is likely to be irreversible. One of the most frightening books written about the galaxy, demonstrating its sheer immensity, is *Powers of Ten*, by Philip and Phylis Morrison. In this illustrated work, we first see a picture of a young couple picnicking in a Chicago park. We turn a page and find a new picture, with its scale reduced by 10. It shows a large part of the city and the couple shrunk almost to a dot. Two more page turnings take us to aircraft altitude of 10 kilometres. Two more, and we are in orbit 1,000 kilometres above the Earth.

This process continues – with progressive changes of scale by a power of 10 – until we are far out in the solar system. The Earth dwindles in size from a large sphere to a small one, then to a blue dot, after which it vanishes. With a few more changes, the transformed spectacle is awesome. We are a light year from Earth, and the Sun is merely the brightest star. At 10 light years, it is no longer brilliant, just an ordinary star, indistinguishable among thousands. It is still visible if one knows where to look. But if a fatal navigational mistake has been made, and one *doesn't*, both the Sun and the crew's destination star will be lost in the immensity of a myriad others, with the vast haze of the Milky Way in the background adding to the confusion. This is the fatal step from which there is no return. Once the crew have lost track of any familiar stars, their predicament will be hopeless. It will be a cold and lonely death.

How does one avoid this peril and navigate one's course safely to a distant star? At first consideration the problem might appear straightforward. What could be simpler, one imagines, than to fix the target star at the centre of the cross-hairs on a forward-facing telescope, reckon speed, position and distance, calculate the time of arrival by dividing distance by intended speed, and sit back to enjoy the ride?

But it is not as easy as that. The first difficulty arises in interplanetary space, before one has even begun to put on speed. The entire appearance of the sky is altered. Faint stars, far off in the galaxy, whose light had been blotted out by the Earth's atmosphere, now shine with brilliant lustre. The familiar shapes of the constellation can no longer be identified with the naked eye. Where, now, is the Big Dipper – sometimes called the Plough – in the constellation of the Great Bear, which was so conveniently used to tell the time in the Middle Ages before the invention of clocks?* (This celestial clock was only accurate to within about

---

* The seven bright stars of the Great Bear which we call the Big Dipper used to be known as Charles's Wain, or Charlemagne's Wagon. Because they never set in

half an hour, and it gained a day every year, but it was good enough for the Middle Ages.)[1] Where are those three familiar stars that form the Summer Triangle, Vega, Deneb, and Altair? How does one find Orion's Belt? All those visions of the night sky that are so commonplace on Earth, even on mountaintops, become submerged in the glittering background.

This first problem will be quite easily solved. What will be obscure to human eyes will be easily discernible with electronic ones. The ship will have an *inertial guidance system*, which I will abbreviate as IGS, like those carried today by submarines and ballistic missiles. These devices keep a past record of all changes in acceleration and all changes of course. Such a system will enable the crew to tell exactly where they are at any given time, and where they are headed. And by calculating how much they have accelerated in how much time, they will discover also how fast they are going.

But the IGS might not be wholly reliable. It could contain an unsuspected bug and misread its data. It could give false information about position or speed without anyone being aware that it was doing so. If this happened the crew could be as good as dead.

In the light of this danger, it cannot be too much stressed how important it will be to *double-check* all possible data about the ship's position. The on-board computers will quickly rediscover the concealed constellations. To back up the IGS information, they will calculate the ship's coordinates by taking the bearings of three stars and measuring the subtended angles between them. Mariners on Earth's oceans who used to steer by the stars needed only two of them to calculate their courses, because the Earth's surface is essentially two-dimensional. But in interstellar space there will have to be *three* such stars because coordinates must be calculated in *three* dimensions.

Consider the desired characteristics of these three 'beacon' stars. Two of them must be very far away, at distances of hundreds of light years from Earth, so that they do not alter their

---

the northern hemisphere but only traverse the sky they were most useful as a clock. 'Heigh-ho!' says a character in Shakespeare's *Henry IV Part I*. 'An't be not four by the day, I'll be hanged. Charles's Wain is over the new chimney, and yet our horse not packed.'

apparent positions in the sky as the ship advances.* The third, by contrast, must be comparatively close by, so that the navigating computer can find its constantly changing position in space by measuring the changing angles between the nearby star and the two much more distant ones.[2]

While the ship is still moving comparatively slowly, the two far-off stars will never seem to change their positions and will act as eternal 'beacons' *because* they are so far away. They will be like the distant mountain peaks from which, before the days of global positioning satellites, yachtsmen used to take their bearings when in sight of shore. To be easily discernible these stars must therefore be what astronomers call 'supergiants', stars that radiate *hundreds of thousands* of times more energy than the Sun does. This is a very rare class of star. None of the stars close to our solar system even remotely resembles this description. Of the 44 stars and binary star systems within 17 light years of the Sun, only four, Sirius, Altair, Procyon, and Alpha Centauri, would be brighter if seen from the same distance.

On the other hand there are rare stars much further away that blaze with extraordinary intensity. These are the supergiants. One such is the red giant Betelgeuse, 600 light years away in the constellation of Orion, lying to the north-east of the famous Belt.† Betelgeuse is a true supergiant. It is the tenth brightest star in the sky – as seen from Earth. It radiates 10,000 times more energy than the Sun, and its diameter is 300 times the Sun's. This means that if it were placed where the Sun is it would swallow up both the Earth and Mars. It is so huge that *38 trillion* – 38,000,000,000,000 – Earths would fit inside it.[3]

Another supergiant star useful for navigation will be Deneb in the constellation of Cygnus, one of the members of the Summer Triangle. Deneb radiates 60,000 times more energy than the Sun, but although it is a brilliant star, we see it as slightly

---

* I give a list of suitable 'beacon stars', with their distances and brightness, in Appendix II.

† The supergiants do not appear to be the brightest stars in the sky, but this is simply because they are so far away. Sirius, the star with the greatest *apparent* brightness, appears so bright because it is only nine light years away. But if Sirius and Betelgeuse were at equal distances from us, Betelgeuse would be 400 times brighter.

fainter than Betelgeuse because of its even greater distance from us, 1,800 light years. Deneb and Betelgeuse are vastly separated in two dimensions, and will be well placed as navigational aids.*

Heading towards the outer regions of the solar system, the ship starts its 100-day acceleration towards a substantial fraction of

* The use of particular supergiant stars for navigation is only a 'temporary' solution to the problem. For in a few million years they will no longer be supergiants! Such monstrous stars as Deneb and Betelgeuse are only monstrous because they are dying. They are putting forth their last bursts of energy before collapsing into white dwarfs, neutron stars, or black holes. And so star-farers millions of years hence will have to find different galactic beacons.

the speed of light. Now an even more bizarre process occurs which, paradoxically, will give the navigator another means of calculating his speed. The stars begin to change their colours. Those in front of the ship turn blue, while those behind become red.

This is caused by a phenomenon called the 'relativistic Doppler effect', in which objects whose light is approaching us shift towards the blue end of the spectrum, while those receding shift towards the red. The Doppler effect, named after the nineteenth-century Austrian physicist Christian Doppler who discovered it, affects the wavelengths of sounds as well as those of light: anyone who has ever listened to a train will know that it makes a higher-pitched sound when approaching than when receding.* It also enables astronomers to study the behaviour of the universe.

This should be explained briefly. Although the furthest galaxies in the universe are enormously distant, many interesting things can be discovered about them from their faint light that reaches us. As the French physicist Armand Fizeau showed in 1848 as he extended Doppler's work, the wavelengths of light emitted by stars (or galaxies) that are moving *away from us* are longer than from those that are stationary. Our eyes interpret these wavelengths as colours. As they grow longer, we see them as a kind of 'cosmic rainbow', for they change in this order: from blue to green, from green to yellow, from yellow to orange, and from orange to red. The *longest* wavelengths are the ones we see as red. Thus, the faster an object is receding from us, the more strongly its colour is 'red-shifted'. The Doppler–Fizeau effect thus provides an infallible guide to what is happening in the real universe. The observation that the colours of the most distant galaxies are most strongly reddened proves that it is expanding.

Conversely, the wavelengths of light emitted by objects approaching us shorten towards the *blue* end of the spectrum. We can thus tell whether a star is approaching us or receding from

---

* The Doppler effect might have been known for centuries but for psychological reasons. As Isaac Asimov once wittily remarked, people who escaped injury after facing a cavalry charge would be so relieved at finding themselves alive that they would not notice the lower note of the receding hoof beats of their enemies.

us, and precisely how fast, by the extent to which its natural colour is accentuated towards the red or the blue. The star Sirius, for example, is receding from us at a moderate pace, while the Sun itself, with all its planets, is moving at about 15 kilometres per second in one particular direction in the constellation of Hercules.[4] How do we know this? Because the colours of certain stars in Hercules are shifted towards the blue.

But 15 kilometres per second is a tiny percentage of the speed of light, 0.005 per cent of it, to be exact. The crew of a starship, when they reach their maximum cruising velocity, will be travelling at 92 per cent of the speed of light, nearly *280,000 kilometres per second*, and at this terrific speed, the Doppler–Fizeau effect becomes extreme. Because the ship is approaching the stars ahead of it, these stars are also approaching the ship, while those behind it are receding. All the stars that lie aft will thus turn dark red, becoming so faint that they will vanish, while those ahead will become a dazzling bright blue.[5]

Not only the stars. There is a phenomenon in space, only discovered in 1965, that had its origin in the Big Bang which created the universe about 15 billion years ago. There is, everywhere, a continuous ripple of radiation, a relic of the Big Bang when temperatures were, for a brief moment, infinitely high. This radiation today keeps all temperatures between the stars at just below three degrees above absolute zero. Using Lord Kelvin's scale of temperatures in which absolute zero is conveniently called zero degrees K instead of the more cumbrous –273 degrees C, this phenomenon is called the '2.7 degree cosmic background radiation'.*

At 92 per cent of the speed of light, the cosmic background radiation will take visible form. It will itself turn blue. The astronauts attempting to navigate the ship will be confronted not only by invisibility behind, so that they cannot see where they have come from, but in front by a blaze of blue light, brighter than the light of the full Moon, making it impossible – visually at

---

* Absolute zero thus remains a purely theoretical concept because it is unattainable and nowhere to be found! The cosmic background radiation affects the entirety of the universe. Only in laboratories investigating the conditions of ultra-extreme cold are temperatures artificially created that are *a few fractions of a degree* above absolute zero. Reaching absolute zero itself is forbidden by the Third Law of Thermodynamics.

least – to tell where they are going.

And the faster they travel, the more this blaze of blue light will shrink in size. At the speed of light itself, if ever that was possible to attain, the entire forward view of the universe would be shrunk into a dimensionless dot almost as bright as the Sun.

How can this be? It is the result of *relativistic stellar aberration*, in which the faster one travels, the more do the positions of the stars appear to change. The astronomer Iain Nicolson gives a good analogy in terms of the appearance of falling rain:

> Suppose you are standing in a shower with the raindrops falling vertically downwards. If you begin to run, the more horizontal the direction of the raindrops becomes. Even if – due to the wind – the raindrops are falling at an angle *from behind*, if you run fast enough they would seem to be falling at an angle from ahead.[6]

It is the same with starlight. To a slight degree, this phenomenon can even be seen on Earth. It is customary to talk about the 'fixed stars' – so little do the stars appear to move. But in reality they are all moving. All, including the Sun, are in constant orbit round the centre of the galaxy. (Many of them, those that are double stars, also orbit each other.) But there are also apparent changes in the positions of stars caused by the Earth's movement round the Sun.*

Stellar aberration was discovered by the eighteenth-century British astronomer James Bradley, who reasoned – and then confirmed by observation – that we never see the stars in their absolutely true positions. Because we are always circling the Sun, starlight must always reach us at an angle. The positions of some stars may thus seem to move through an ellipse during the course of a year by up to just over 1 per cent of the size of the full Moon.[7]

---

* This is a much greater effect that the stellar *parallax* that enabled Thomas Henderson to discover the distance to Alpha Centauri, in which the nearer stars appear at different positions at varying times of the year as the Earth moves through its 300-million-kilometre-wide orbit.

The speed of the Earth's movement round the Sun, however, is only 30 kilometres per second, 0.01 per cent of the speed of light. A starship, by contrast, will be moving at 92 per cent of the speed of light. The faster one travels, the more the stars seem to change their positions. At the starship's cruising velocity, some of the stars will have clustered in front of the ship.[8] The giant 'beacon stars', so useful for navigation at low speeds, will have vanished among them.

But this extraordinary forward view from the ship, as if the stars were clustering to obstruct its passage, will do more than merely bewilder first-time star travellers. Weird though the spectacle may appear, it will supplement the IGS in providing *an accurate interstellar speedometer*. It will enable the navigator to recheck another vital piece of information.

Earlier, when plotting the positions of the three beacon stars, he needed to have observed a fourth star. So as not to confuse it with the beacon stars, I will call this the 'guide star'. It must lie exactly at a right angle to the ship's direction of motion when the vessel is moving comparatively slowly. At a slow speed, therefore, the position of ship and guide star would appear as figure *(a)* below.

But when the ship accelerates to its cruising velocity of 92 per cent of the speed of light, the position of the guide star, as seen from the ship, will have changed. Moving towards the deepening cluster in front of the ship, it will appear as figure *(b)* below.

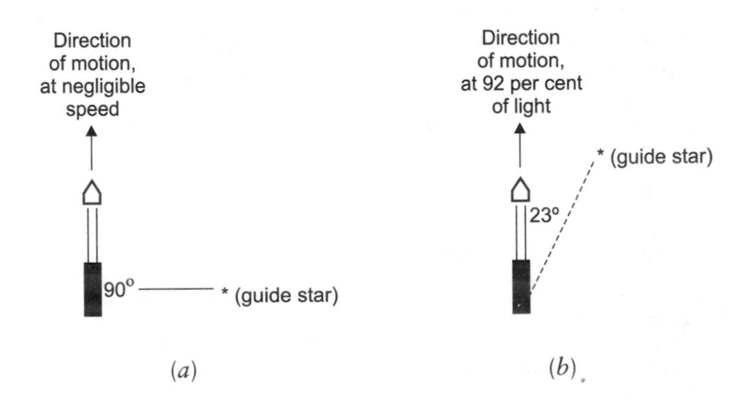

(a) *(b)*

Here is a table showing how the ship's speed will bring about a corresponding apparent displacement in the angle of the guide star:[9]

Ship speed as % of light	*Displacement of guide star (degrees)*
0	90
10	84
20	78
30	73
40	66
50	60
60	53
70	46
80	37
90	26
92	**23**
95	18
99	8

'How fast are we going?' is now an easy question to answer. The navigator need only look at the above table, which will appear on his computer screen, to see the new position of the guide star. If all is going as it should, the figure of 23 will be highlighted, as it is in the table. A glance at the left-hand column tells him his speed as a percentage of the speed of light. From the IGS he will know his course, and he will know for how long he has travelled. His speed now tells him the remaining distance to the target star. He divides it by his future intended speed and he can estimate his time of arrival.*

Seeing all the stars change their colours and their positions in

* The crew might have an unpleasant shock if their estimate of speed derived from the angle of the guide star *disagreed* with that from the IGS. It seems obvious that, as far as *speed* is concerned, the former is much more likely to be accurate than the latter, being based on a simple observation rather than on a complex electronic instrument. But if the two *speed* estimates differ, what, then, is the ship's *position*? If the IGS has misinformed them about their speed, then it will certainly be misinforming them about their position. They could take the precaution of having *two*, independently functioning IGS machines. But what if *both* these systems gave contradictory data? One must face the fact that there will always be risk and that not even multiple systems of navigation will be 100 per cent safe.

so strange a fashion will be an extraordinary visual experience. It may introduce in the crew a kind of 'cosmic claustrophobia'.[10] With the stars turning blue and clustering together, and the background relics of the long-ago formation of the universe itself taking on visual form, it will be a remarkable transformation as the ship speeds up, which will vanish again as it slows down. As two scientists describe these effects:

> Ultimately, at the very highest attainable speeds, the all-pervading 2.7 degree K cosmic background radiation would be blue-shifted into the visible region and reach its brightest intensity in a forward cone spanning less than one minute of arc [one thirtieth of the diameter of the full Moon], with a visual magnitude equivalent to about one tenth of the brightness of the Sun as seen from Earth, *with the remainder of the sky completely black* [my italics].
>
> The view of the universe in this extreme would thus dwindle to a mere pinpoint of light, having dazzling brilliance with its surround of utter darkness. Only the spacecraft and its occupants would retain the familiar span of space and time. Upon deceleration, environmental space and time would revert to being in synchrony with that experienced by the astronauts. The initially all-embracing universe would have seemed in the course of a single flight to have become crowded forward into a brilliant minute lodepoint of light and then, like an unfolding rosebud, to have opened up again into its full majestic breadth and 'natural' colour.[11]

The Ancient Mariner, finding the nocturnal sea blazing with electrical effects that he did not understand, was completely lost and disoriented. Not so the starship crew. They will find their surroundings strange, but not impossible to comprehend. And unlike the Ancient Mariner, they will be able to make practical use of them. As Eugene Mallove and Gregory Matloff point out in their book *The Starflight Handbook*, 'if reactions to past supposed barriers are any guide, "high-speed" travel by rail in the era of horse transport, supersonic flight, and weightlessness, human beings will adapt with ease to the bizarre vista from the relativistic starship bridge'.[12]

Chapter 13
Faster Than Light?

They made themselves air, into which they vanished.

William Shakespeare, *Macbeth*

Any sufficiently advanced technology is indistinguishable from magic.

Arthur C. Clarke's Third Law

There is no question that there is an unseen world. The problem is: how far is it from mid-town and how late is it open?

Woody Allen

How convenient the future would be if the speed of light was not a universal speed limit, if spaceships could travel as fast as their owners liked! The social implications would be tremendous. We would be living in a galaxy (never mind the rest of the universe) where it would take no more than a few months at most to travel between one star and another, even between stars at opposite ends of the galaxy.

It would be possible, in this situation, to turn all the galaxy's probable tens of millions of habitable worlds into a single political union. (We'd be back with those politicians again.) A Galactic Empire could exist, as in Isaac Asimov's *The Foundation Trilogy*. Messages would be transmitted instantaneously between the centre of the empire and the outlying provinces instead of taking thousands of years to arrive, and people doing a future equivalent of the Grand Tour could visit hundreds of different planets during their lifetimes, even those separated by thousands of light years.

In most science fiction, this kind of universe is standard, since if people can interact across a galaxy there is much more scope for melodrama and politics than if they are confined to isolated groups of planetary systems. (Imagine *Star Wars* without its Emperor and its rebels!) But in reality, on a first consideration at least, it seems very unlikely that people could travel in this fashion. Einstein's 1905 special theory of relativity, which restricts the speeds of all material objects to below the speed of light, and forbids any signal from exceeding it, has been confirmed by all tests. When matter has been accelerated in atom-smashing machines to close to the speed of light, its mass increases exactly as Einstein predicted. The inference is obvious. As I pointed out earlier, the mass of a spaceship travelling at just below the speed of light would be almost infinite, and so to accelerate any further would require an infinitely powerful engine. Since no such engine can ever exist, faster-than-light travel must be impossible.

Science fiction writers – the best of them, anyway – admit the truth of this argument. But while sticking as close to science as they can, they try to find another way round it. Their spaceships 'jump' or go through 'star gates'. They vanish in one region of space and reappear an instant later in another. To do this, they go through a mysterious dimension called 'hyperspace' whose physical laws are different from those of normal space. Asimov, in his novel *The Stars Like Dust*, described such a journey with his customary clarity. A spaceliner is leaving Earth for a distant star, and its passengers hear the announcement:

> 'This is the captain speaking. We are ready for our first jump. We will be temporarily leaving the space-time fabric to enter the little-known realm of Hyperspace, where time and distance have no meaning. It is like travelling across a narrow isthmus from one ocean to another, rather than circling a continent to accomplish the same distance. There will only be minor discomfort. Please remain calm.'
>
> It was like a bump which joggled the deep inside of a man's bones. In a fraction of a second the star view from the portholes had changed radically. The centre of the great Galaxy was closer now, and the stars appeared to thicken in number. The ship had moved a hundred light years closer to them.[1]

This was a fictional exercise in what physicists call 'crazy ideas' at the cutting edge of science. There is just a chance that they may contain some truth that starships may be able to exploit, and no book on this subject would be complete without a discussion of them.* Of necessity, therefore, this chapter must be somewhat more speculative than the others. Note the question mark in its title.

There is a saying: 'We broke the sound barrier, so why can't we break the light barrier?' But achieving the latter appears immeasurably more difficult than the former. Nevertheless, let us try.

It is less widely known that in 1916 Einstein published a second theory of relativity, the 'general theory', which is much more subtle and complex than the special theory. It states that matter creates the space and time which surround it. This in turn indicates that space has an elastic, or warped character. A large mass, such as a star, causes space to curve in the region immediately surrounding it.†

But this is not all! Einstein's 1916 equations have been compared with a Trojan Horse. Ostensibly, there is little more in them than a description of light and space being bent by gravity. But in fact, as one physicist has remarked, within them lurk all sorts of strange 'goblins and demons', black holes, time travel and other universes.[2] Small wonder that the relativistic cosmologist Kip Thorne subtitled one of his books 'Einstein's Outrageous Legacy'. In seeking ways to achieve instantaneous flight, we shall encounter these bizarre phenomena.

The curvature of space around stars is only an extreme case of a much smaller *net curvature* that exists everywhere. A simple experiment explains this idea. Imagine space like a flat piece of paper. Draw two dots on it to indicate two stars many light years apart. Now fold the paper so that the two dots touch. They are

---

* Einstein himself, after his life's work was complete, was at pains to stress how little he had discovered. He told an interviewer that he thought of himself as a child who has entered a huge library filled with books written in many languages. He has taken down a single volume, of which he has only managed to translate a few pages.

† When space is curved, the path of light rays through it also becomes curved. The truth of the general theory was confirmed in this way by an experiment in 1919. During an eclipse of the Sun (when it was possible to see daytime stars), a star was seen to have apparently moved from its normal position by 1.7 seconds of arc. Its light path, when passing the Sun, had been curved by the Sun's gravity.

still the same distance apart in normal space – the paper is still the same size – but there is now a short cut between them whose distance is zero.

The general theory shows that this is what space is really like. Down at the sub-microscopic level, it has a 'foam-like' structure. At the scale of the Planck constant, the smallest distance that can exist, a billionth of a trillionth of a trillionth of a centimetre (10^{-33} centimetres), it is pierced with tunnels or 'wormholes' like a Swiss cheese. These interconnect parts of normal space through the dimensionless region called hyperspace. To emphasise this point, Einstein and his colleague Nathan Rosen published a paper in 1935 showing how distant points in space could indeed be connected by a 'bridge', a concept since known as an Einstein–Rosen Bridge.[3] As for the true nature of these wormholes or bridges, the physicist John A. Wheeler gave this description in 1963:

> Space is like an ocean which looks flat to an aviator who flies high above it, but which is a tossing turmoil to the hapless butterfly fallen upon it. Regarded more and more closely, it shows more and more agitation until . . . at distances of the order of 10^{-33} centimetres the entire structure is permeated every which way with wormholes. If geometrodynamic law is correct at such distances, it forces on all space this foam-like character.[4]

This sounds all very well, but from the practical point of view of sending spaceships through these holes, there does seem to be a certain problem of scale. The width of a wormhole, 10^{-33} centimetres, might create rather cramped conditions for a ship containing 900 or so people! This is not only 33 powers of 10 smaller than a human thumbnail; it is also 20 powers of 10 smaller than an atom. When Asimov shrank a manned vehicle down to the size of a virus (about 10^{-8} centimetres) in his novel *Fantastic Voyage*, he made it sound scientifically plausible (well, almost), but the prospect of shrinkage on *this* scale sounds utterly inconceivable.

So the matter rested until the 1980s when Carl Sagan was doing research for his novel *Contact*. He needed a plausible way to get one of his characters quickly to another star without worrying about that irritating speed-of-light restriction. He consulted Kip

Thorne. Thorne had long dismissed the feasibility of hyperspace journeys, but he came to change his mind. With his colleague Michael Morris, he published a paper showing how such a journey could in principle be made.[5] The question, as Thorne put it, to emphasise its purely theoretical nature, was:

What things do the laws of nature permit an infinitely advanced civilisation to do, and what things do those laws forbid?[6]

Their solution, the answer to this stupendous question, was not to shrink the spaceship, but to enlarge the width of the hole, as one sometimes has to enlarge a brand-new pair of gloves by forcing one's fingers into them. The key is that wormholes have a positive electric charge, and positive charges repel one another. A sufficiently strong electric field, therefore, forced down the wormhole, could open it until it was wide enough to admit a spaceship. This could be achieved by building two gigantic metal spheres, both of them perfect conductors of electricity, and placing them several light years apart in space. The electric field generated by each of them would keep the tunnel open.[7]

There are two snags. To generate a sufficiently strong electric field, each sphere would have to be some *70 million kilometres wide*, half the distance between the Earth and the Sun, which raises the spectacle of enormous difficulties and costs. And while one sphere would remain fairly near Earth, the other, destined for the opposite end of the wormhole, would have to be propelled there through normal space since it would not have any wormhole to travel through. And if the intended wormhole was to carry astronauts to a distant star system, the second sphere might take a frustratingly long time to get there and start its work.

One is reminded in this context of Guglielmo Marconi, who in 1895 proposed that radio messages could cross the Atlantic. Critics laughed at his suggestion, saying that to do this he would need a transmitter as large as the North American continent.[8] But in an obscure sense his critics were right. He didn't *need* a transmitter as large as this, but if he had *had* one it no doubt would have transmitted the radio messages. If it is shown that ludicrously inappropriate machinery can solve a problem, then it is shown in turn that the problem is soluble.

Of course, if mankind's instantaneous journeys were going to be confined to between Earth and *one* other star system, the construction of these giant spheres might be justified. But the sphere near Earth would create environmental problems – its mass would distort the orbits of planets anywhere near it, to say the least – and its builders would face constant carping to the effect that there must be a cheaper and simpler way to achieve the same result.

Indeed, there probably is. Thorne and Morris were only doing the best they could within the time and knowledge available to them. (Sagan had a deadline for his novel.) Moreover, the general theory of relativity has only been public knowledge for less than a century – although it has been operational for some 15 billion years – and we are far from having explored all its ramifications and possibilities.* As the physicist Frank Tipler remarks, 'solutions to its field equations can be found which exhibit virtually any type of bizarre behaviour'.[9]

There may be another way to expand the passage of the wormhole, although nobody at present knows what it could be. We only know that *in principle* it could be done, albeit at vast and unacceptable expense.

And there is another frightening problem. Where would a wormhole lead to? In the present state of thinking, entering one would be like going blindfold into a subway system. You would have no idea where you were going to emerge, if indeed you were going to emerge at all. A brilliant technology might be created for instantaneous interstellar travel, the sole effect of which might be to make people vanish for ever. Victor Weisskopf, a theoretical physicist at MIT, once remarked that his fellow physicists are divided into three kinds, who do not always communicate with each other very effectively. He compared them with the three classes of experts who managed Columbus's voyage of 1492 which landed him unintentionally in America:

> The machine builders correspond to the captains and ship builders who really developed the techniques at that time.

---

* This may seem an astonishing statement when one reflects that the original document, Einstein's *Die Grundlage der allgemeinen Relativitätstheorie* ('The Foundation of the General Theory of Relativity'), published by the journal *Annalen der Physik* (Vol. 49) in 1916, fills a mere 55 pages! But it is packed with the highly compressed language of tensor equations. To put it mildly, it was capable of expansion. In 1973, Misner, Thorne, and Wheeler published their 1,277-page *Gravitation*, a book which explores similar ground.

> The experimentalists were those on the ships that sailed to the other side of the world, and then jumped upon the new islands and just wrote down what they saw. The theoretical physicists are those fellows who stayed back in Madrid and told Columbus that he was going to land in India.[10]

If you are seeking a route to India to revolutionise the spice trade, you don't *want* to end up in a mysterious place in between and not have any idea where you are. Your voyage might be epochal, but from the viewpoint of your original purpose it is useless.

For while Einstein's bridges seem to exist, they may not necessarily lead to where we want them to. There are those who believe that wormholes are not gateways to other regions in space, but paths into other universes from which there might be no escape. The theory, very much simplified, is this. When the universe was first created in the Big Bang, it was no larger than the mouth of a wormhole but infinitely dense. Then it underwent 'inflation'. Within a fraction of a second it expanded many times faster than light, reaching something like its present size. But during this inflation it divided and subdivided many times, creating innumerable 'bubbles'.[11] Each bubble went its own way as a different universe, and ours is one of them.*

These universes would be liable to have different physical laws from each other. Some might be made of antimatter, so that entering them in a ship made of matter would be an explosive experience. Some would be cold, dark and empty of anything solid, because gravity would be too weak to allow stars to form, and there would consequently be no elements heavier than hydrogen and helium. In others, the speed of light might be instantaneous (now *that* would be a convenient universe to find oneself in!). And there could conceivably be some in which gravity was reversed, so that water ran uphill and people needed roofs over their streets to stop them from flying up into the sky. Tipler was certainly right about Einstein's equations permitting bizarre behaviour.

---

* In the light of these beliefs, it has been proposed that the word 'universe', when it means the entirety of everything that exists, should be changed to something more appropriate for many universes, like the 'multiverse'. But the suggestion has not generally caught on.

Most important, there would certainly be some universes in which time ran backwards. There is a principle in physics called the 'arrow of time'. This does not state that time in a universe must always flow forwards, as it does in ours, but only that it must always flow in the same direction.[12] Astronauts who entered a universe in which time ran backwards would find themselves in a time machine. *Their* time would run backwards like that of the universe they had entered. It is very hard to visualise such universes, and the best descriptions that have been given of them are in the form of jokes. This does not invalidate the idea; it only shows how difficult it is for us to visualise them. John D. Barrow, in his book *Impossibility*, quotes from a story called *The Oscillating Universe* by Dennis Piper:

> One day the Professor called me in to his laboratory. 'At last I have solved the equation,' he said. 'Time is a field. I have made this machine which reverses this field. Look! I press this switch and time will run backwards run will time and switch this press I. Look! Field this reverses which machine this made have I. Field a is time,' said he, 'Equation the solved have I last at.' Laboratory his to in me called Professor the day one. 'For heaven's sake. SWITCH IT BACK,' I shouted. *Click*! Shouted I, 'BACK IT SWITCH, sake heaven's for.' One day the Professor called me into his laboratory . . .[13]

In principle, these astronauts could return to *our* universe and be in *our* past. They would be in a position, if they chose, to alter *our* present.

But it is very difficult to imagine this situation. Someone who travels back in time and kills his parents before they met not only ceases to exist but *ceases ever to have existed*. Science fiction writers have had great fun with this idea, but it remains an impossible paradox. Indeed, Michio Kaku, in his book *Hyperspace*, points out that backward time travel (there is no problem with *forward* time travel; see Chapter 7 'The Incredible Shrinking Calendar') would turn reality into chaos:

> Time travel would mean that no historical event could ever be completely resolved. History books could never be written. Some die-hard would always be trying to assassinate

> General Ulysses S. Grant or give the secret of the atomic bomb to the Germans in the 1930s.
>
> What would happen if history could be rewritten as easily as erasing a blackboard? Our past would be like the shifting sands at the seashore, constantly blown this way or that by the slightest breeze. History would be constantly changing every time someone spun the dial of a time machine and blundered his or her way into the past. History, as we know it, would be impossible. It would cease to exist.[14]

We learn a great deal from observation that history *isn't* like this. The fact that backward time travel *does* not happen in our universe is evidence that it *cannot* happen.* If it were possible, people in our future would have invented it and indulged in it. As Stephen Hawking believes, it is proved to be impossible 'by the fact that we have not been invaded by hordes of tourists from the future'.[15]

However, we are still faced with the arrow of time, and another principle of physics which states that anything which *can* happen *does* happen. Suppose an astronaut goes to one of these backward-time-running universes, and then returns to our past. He might be able to do so, but only as an invisible observer. The experience would be like watching a newsreel of an old battle and being unable to interfere with its outcome. In this way, no paradoxes are created and history is not violated.

But – and this is the final 'but' – this idea does not work. No conscious being could ever be restricted to the status of an invisible observer. He or she would always be able to do things, even the smallest things, that would influence the future. And yet, as we have seen, this does not happen! The past does not change. The only possible answer to this riddle, proposed by David Deutsch, is that astronauts who imagined that they had returned to the past of their own universe *would not in fact have done so* – only to the past of a universe that is very similar to it. Once you have entered the past of our universe, it immediately switches itself into another. You can change *its* past and present, but not *our* past and *our* present.† By this means, nature prevents its own laws from being violated. It sounds a daring, even a crazy solution, but it is the best that physicists have so far come up with.[16] In short, once you have left this universe, you are gone for ever. Universe-hopping is a one-way journey. We are back where we started, lost in other universes amidst a labyrinth of wormholes.

---

* There is another way of looking at this argument. If history was being constantly altered, we might not be aware of it. Each change of reality would create a new collection of memories and a new set of history books. They would appear as if by magic and we would 'know' that they had always been there. But this idea cannot be tested, so perhaps there is no point in pursuing it.

† According to one bizarre possibility, someone who entered another universe would never know that he had done so. For he would instantly acquire a new and presumably fictitious memory. As Sir Fred Hoyle puts it, 'there is the possibility of waking each morning beside a different spouse, although our memory each morning would always be consistent with the spouse-of-the-day, and we will therefore be entirely unaware of the other possibilities'.

★ ★ ★

In 1994, NASA's Office of Advanced Concepts and Technology sponsored a two-day meeting in Pasadena, California, to discuss possible means of achieving faster-than-light travel. The space agency held a second such meeting at its Lewis Research Centre in Cleveland in 1997. Both these seminars were attended by many eminent physicists and hard-core science fiction writers.

Many and varied were the topics raised, both informally and during the public discussions.[17] Wormholes and other universes, and time travel paradoxes were talked about exhaustively. So were 'tachyons', theoretical particles – the name is based on a Greek word meaning swift – that always travel faster than light and never slow down. Tachyons, if they exist, would not violate the special theory of relativity, which only restricts moving objects from actually *crossing* the light barrier. It permits speeds on either side of the barrier.[18] Might it not be possible to hitch a ride on a tachyon? Or use it as some kind of propellant? At least one science fiction novel has a starship with a 'tachyonic drive'. The trouble in the real world is that exploiting tachyons might be like trying to board an aircraft that never lands. If it never slows down to below the speed of light, how would one rendezvous with it? Moreover, an object that travels faster than light travels backwards in time, so people who travel forward in time, like us, would be doubly hard-pressed to make contact with it. But participants vowed to keep the idea in mind.

There are still more germs of interesting ideas. Arthur C. Clarke talked in a short story of future starships that 'eat up the light-years'.[19] This, essentially, is what the physicist Miguel Alcubierre's proposed 'warp drive' does. On a local scale, it imitates the inflation that followed the Big Bang. It causes space behind the ship to increase in volume and space in front of it to shrink. In other words, the ship does not move, but the region of the universe that surrounds it does. It would be like a car that remained stationary while the road ran backward beneath it.[20]

Unfortunately, while Alcubierre's idea is fully consistent with the general theory of relativity, it requires extraordinary technologies. It needs exotic matter with negative energy, and physicists are not agreed about what this concept means, let alone how to create it. To move even a local region of the universe in this fashion may require the influence of mass on a planetary scale, like Thorne and Morris's exercise in opening the wormhole passage. Another physicist drily suggests that an Alcubierre journey would require an environmental impact statement. Planets and other spaceships that got in the way of his ship would be enveloped and destroyed by the local distortion of space that it created.[21]

A hyperspace universe would also bring a tremendous advantage not only in travel itself, but in *navigation*. It would be possible to have variable charts of the galaxy showing how the stars would look from every different point in space, depending on where you were.

As we saw in the last chapter, this cannot be done in a universe where all speeds are restricted to below that of light. For as soon as a significant percentage of the speed of light is reached, the appearance of the cosmos is changed by the effects of special relativity, and any such charts would become useless at the very moment when they were most needed.

But in a hyperspace universe, no ship would ever have to travel very fast. It would merely have to go fast enough to reach within a reasonable voyage time the point where it is going to 'jump' into hyperspace. This would be well below 1 per cent of the speed of light, and at this speed the appearance of the universe would be completely normal.

One can therefore envisage – in a hyperspace universe – computer software packages that would give a complete simulation of the appearance of the local stellar neighbourhood as seen from any given region. (The night sky would look notably different, for example, from a planet in Alpha Centauri than it does from Earth.) Before entering hyperspace, and after emerging from it, the navigator would punch in his three-dimensional coordinates, and a view of what he *ought* to be seeing would appear on his screen. If it corresponded to what he could see from the forward view on the flight deck, he would know he was in the right place. If it did not correspond, he could presumably, by some electronic trick, describe to the computer what he could see from the flight deck. The machine would identify this place from its gigantic database, and he would then be able to discover where he actually was.

It might be objected that constructing such software would require a prodigious amount of computer memory, far more than can be found in today's most powerful computers, and perhaps beyond human capability. Vast indeed would be the data describing a galaxy of hundreds of billions of stars, informing us not only what space would look like from every one of them, but from every point *between* them! But this difficulty is less great than it might appear. We would only want information that was useful, only of the regions around the stars that had already been colonised or surveyed by probes. The scope of the charts would gradually expand with interstellar migration.

We already possess excellent star charts, in both printed and electronic form, showing details of star fields in ever-improving quality. But they only show this detail *as seen from Earth*. It is true that with an electronic star chart one can 'zoom in' and get an expanded view of a star field. But one cannot, from this vantage point, 'turn around' and look back at the solar system from afar. A true galactic chart, on the other hand, would be the nearest interstellar equivalent to a terrestrial atlas, where one finds one's way merely by choosing the map of any desired country.*

What, then, are the chances that a starship captain, aided by such a device and by hyperspace, will one day be in a position to tell his passengers: 'We are ready for our first jump'? At this stage, the answer depends on one's philosophy of life rather than on any reasoning. If one is an optimist, one cannot help believing that the captain's announcement will be made. Our knowledge of physics is proceeding at an exponential rate, doubling every few years. We learned more during the twentieth century than during the previous 1,000 centuries. If there *is* a way to go through wormholes or any other kind of holes without dismantling the solar system, then it will be found, and barring any opposition from aliens, human galactic empires will be created.†

On the other hand, some people argue that we have reached the end of science. However much knowledge we have acquired, so this argument goes, there now isn't any more to be found. We

* The 'inventor' of such a chart was Asimov, in his novel *Second Foundation*. He called it the Lens, the 'newest feature of interstellar cruisers'.

† In my super-optimistic younger days (1977), I wrote a book called *The Iron Sun: Crossing the Universe Through Black Holes*. It now appears that the infinite gravity at the heart of a stellar-sized black hole (the giant cousin of a wormhole) would crush any spaceship that attempted to fly through it. But scientific opinion may change . . .

know the universe more or less as well as we are ever going to know it.[22] If this applies to hyperspace, then farewell galactic empires. Instead, our descendants will flourish in an unlimited number of *local* empires, isolated from each other by the speed-of-light barrier.

And what difference will it make? I have made a rough calculation that is far from rigorous – a 'guesstimate' would be a good word for it – about the timescales involved in colonising the entire Milky Way galaxy depending on which of these alternate technologies is used.

Recall that the galaxy is about 100,000 light years across, and that it contains about 250 billion stars. Being reasonably conservative about the number of suitable stars, this would give us, perhaps, 30 million or so habitable planets. My estimate, for what it's worth – and assuming our path is not blacked by alien civilisations – is that *with* hyperspace colonisation will take about 40,000 years, and *without* it, about five million years. In other words, hyperspace would speed up the rate of colonisation more than a hundredfold. Whether it will ever become possible and practical, however, still depends on the unknown.

Chapter 14

'I Enjoy Working With People'

The stars were as thick as weeds in an unkept field, and for the first time Lathan Devers found the figures to the right of the decimal point of prime importance in calculating . . . There was a frightening harshness about a sky which glittered unbrokenly in every direction. It was like being lost in a sea of radiation.

Isaac Asimov, *The Foundation Trilogy*

The first super-intelligent machine that man invents will be the last invention he will be allowed to make.

Arthur C. Clarke, many years before inventing HAL

The year 1997 is supposed to have marked the birthday of a very important character in science fiction. HAL, the all-intelligent – but murderous – computer in the film *2001: A Space Odyssey*, will be remembered when countless film dramas about space travel of lesser quality have been forgotten.* In fact, the more often I see *2001* – and other space dramas with strong

---

* HAL stands for 'Heuristically Programmed Algorithmic', *not* IBM with each letter put back a space in the alphabet, as has frequently been claimed. Clarke's novel *2010* contains this dialogue between HAL's Earth-bound counterpart SAL and its chief programmer:

'Is it true, Dr Chandra, that you chose the name HAL to be one step ahead of IBM?'

'Utter nonsense! Half of us come from IBM and we've been trying to stamp out that story for years.'

scientific backgrounds like *Alien* and *Aliens* – the more interesting I find them; whereas the more often I see such films with minimal science, like *Star Trek* and the *Star Wars* series, the more ridiculous I find them.

'I became operational at the HAL plant in Urbana, Illinois, on January 12 1997,' says HAL in Arthur C. Clarke's novel of this film, a date that was far too early to be realistic, for no one knows yet how to design a machine even remotely as powerful as HAL. Nevertheless, it appears self-evident that a voyage to another star will be so lengthy, so hazardous and so complex that a computer of barely imaginable power, speed, and capacity will be needed to watch over it. But we are still far from building such a super-intelligent machine. Indeed, as one astronomer has pointed out, perhaps with some exaggeration, 'it is a deficiency in computer technology, not in rocket technology, which prevents us from beginning the exploration of the galaxy tomorrow'.[1]

On a much smaller scale, transport vehicles are being similarly computerised today. The most modern jet aircraft such as the Boeing 777s and the Airbus 340s are almost entirely flown by computers. But these machines which enable the vehicle to 'fly by wire' are essentially 'dumb'. They are only programmed – indeed they only *can* be programmed – to handle the very narrow set of conditions that can arise during a flight through the atmosphere. For a starship, a very different kind of computer will be required.

A single example of the powers it will need involves the tremendous difficulties of super-accurate navigation where a single 'down' period, even if it lasts no more than an hour, could be catastrophic. As explained in Chapter 12, 'The Star on the Starboard Beam', it will have to guide the ship through vast, empty regions where, since one direction in space can look indistinguishable from another, there will be no outward clue to one's position. Every light year that the ship moves, the familiar shapes of the constellations will become distorted until they are utterly strange. If we were to look back from Alpha Centauri, for example, the Sun would appear at the edge of the constellation of Cassiopeia.[2] But if a navigational error had been made and the ship was off course, it might be impossible to find the Sun, since *it would be projected against a background of stars that was entirely different*. The error would probably be fatal to the mission. The crew would have no idea where they were and no means of finding out. Barring the extremely unlikely event of finding a habitable planet by accident, all aboard would perish.

It is idle to imagine that any immediate successor to today's most advanced personal computers, or even modern supercomputers, could perform such tasks, requiring calculations millions of times faster than present-day machines are capable of, with an accuracy of at least 50 decimal places.

HAL, the most famous computer in science fiction, is the 'Mecca' of all computer scientists who dream of constructing a super-intelligent machine. A common topic of fascination is how near we are to building such a device. I make no apology for writing about HAL, since almost everyone interested in futuristic science and technology has seen the film, which, although released back in 1968, remains a classic. HAL can handle any crisis that might affect the ship. He can also talk, recognise faces, understand human speech, and look after the health of the astronauts who are in hibernation. But there is a dark side to his nature. Despite all his smooth and friendly utterances, he has no moral sense. He suffers from what psychologists call the 'Byzantine generals' problem' – the inability to behave in a reasonable manner when he can no longer conceal the fact that he is acting under contradictory orders.[3] He has been ordered by Mission Control to keep secret from the crew the purpose of the mission. But the crew must be informed of this secret if they are to carry out the mission. He *must* inform them, but he *cannot* inform them. Round and round, in his electronic mind, the circle goes. His final solution to the riddle: to kill the crew and try to carry out the mission alone.

A computer in charge of a starship *must* be given some of the attributes of HAL if it is to perform its central task, to ensure that the ship and crew arrive safely at their chosen destination. Literally, one could call these attributes a 'sense of priorities'.

Another way of saying this is that to do his job, HAL must be emotional.

Any computer that is to be called 'intelligent' must, in this sense, be emotional. Otherwise, despite its tremendous speed at processing information, it is just a stupid slave like a computer today. A machine with a 'mind' powerful enough to qualify it for supervisory tasks in a starship must have this quality – or at least act as if it does.* Its emotions must enable it to behave sensibly in a crisis. When it is confronted by two or more alternative courses of action, *any* of which could be disastrous, it must choose what it believes to be least disastrous.

Some computer scientists are now trying to teach computers to drive cars.† Such a computer will face dangerous choices, as human drivers constantly do. It will therefore need a set of rules such as these:

1. It is worse to kill a child than to kill a cat.
2. It is worse to kill a cat than to hit another car's bumper.
3. It is worse to hit another car's bumper than to park illegally.

Even playing games requires critical decisions. One of the most dramatic scenes in the film is when HAL plays chess with the astronaut Frank Poole (whom he afterwards murders). It only lasts 30 seconds, but for what it reveals about HAL's mentality it is remarkably subtle. We are briefly shown this chess position:

White (Poole) is about to play his 13th move, and his position is hopeless. I have played this position several times on a chess computer, with myself as White, and there is no way to avoid checkmate in four moves at the most so long as Black plays above the level of a novice. In this position, while Poole's Queen is beleaguered at the far end of the board, a formidable phalanx of Black officers threatens his King.[4] The dialogue between Poole and HAL goes like this:

---

* As one of the astronauts says in the film: 'HAL acts like he has genuine emotions. Of course he's programmed that way to make it easier for us to talk to him. But whether or not he has real feelings is something I do not think anyone can truly answer.'

† One such is Dean Pomerleau of the Robotics Institute at Carnegie Mellon Institute in Pittsburgh, whose laptop computer successfully drove a Pontiac mini-van 400 kilometres by motorway from Pittsburgh to Washington. The laptop took charge of the steering, braking, and acceleration, but with constant supervision from Pomerleau. The machine's only fault was, repeatedly and inexplicably, to turn the vehicle into exit side roads without being told to. 'Cars that drive themselves so safely that you can tell them your desired destination and go to sleep are still at least 10 years away,' Pomerleau reported after the experiment in 1995.

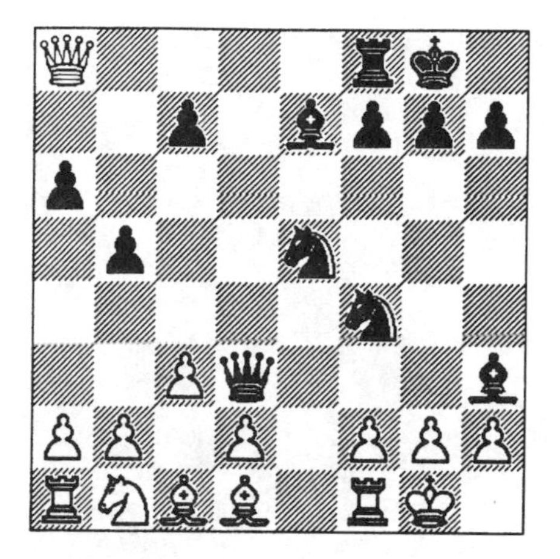

POOLE: Umm. Anyway, Queen takes Pawn.
HAL: Bishop takes Knight's Pawn.
POOLE: Lovely move. Er . . . Rook to King One.
HAL: I'm sorry, Frank. I think you missed it. Queen to Bishop Three. Bishop takes Queen. Knight takes Bishop. Mate.
POOLE: Ah, yeah. [*After a very brief pause.*] Looks like you're right. I resign.
HAL: Thank you for an enjoyable game.
POOLE: Yeah. Thank you.[5]

The biggest surprise to me was the discovery by a computer chess expert, Murray S. Campbell of IBM, from the brief glimpse of the board shown in the film, that this was the end of an actual game.[6] It was not a random chess position, as producers of a second-rate movie might have created. The game whose end we see in *2001* was from a tournament played in Hamburg in 1913. The standard of play is mediocre, which may explain why the game does not appear in any of the great chess databases. In this game Black (HAL) does not play like a machine at all. *He plays like a human.*

I know of two ways to play chess. One, known as the 'Minimax method', practised by all serious players, is to try to make the best possible moves – as a chess computer does – and to assume that your opponent will do the same. The other is to

adapt your play to the personality and mental limitations of your opponent.*

Being a poor player, I always choose the latter. I have as my favourite chess opponent a retired diplomat who is somewhat emotional and immodest and who likes a glass at his elbow. On the rare occasions that he defeats me, he jumps up delightedly and does war-whoops. I know of one infallible defence against him.

---

* Stephen Potter, in his amusing 1947 book *Gamesmanship: The Art of Winning Games Without Actually Cheating*, suggests ways of playing in this fashion. With an aggressive opponent, move your pieces soundlessly and delicately; this is bound to irritate. But when playing against a timid person, bang down your pieces noisily and forcefully. It can also be useful to make patronising remarks about your opponent's play, like: 'Are you sure you meant that?' or: 'Your castle won't like that in six moves' time.' Another off-putting ploy is to say: 'Hmm, an interesting position. Do you mind if I make a note of it? The *Chess News* usually publishes any stuff I send them.'

When in a tight spot, I say: 'My dear fellow, have some more whisky.' He accepts and his standard of play deteriorates.

HAL, in the 1913/*2001* game, is not as crude as this, but he does play in the clear knowledge that Poole is a weak opponent. He treats Poole almost with contempt. He sets obvious traps for him that a strong opponent would never fall into. He plays very aggressively and takes risks. He even overlooks an opportunity to win a move earlier than he does.* Poole knows that HAL is a far better player than he is. (Being a professional astronaut, he would have little time to practise.) Hence Poole's very brief pause before announcing his resignation.[7] He does not even trouble to check HAL's reasoning about the position. HAL seems anxious to win as quickly as possible – perhaps he is anxious to get on with his murder plot.

This subtlety in HAL's characterisation (although hidden to all but chess experts like Campbell) is a reason why, to prophets of future space travel, *2001* is such an extraordinarily intelligent film. HAL, although a computer, has uncannily human-like emotions. I believe him when he says the game has been 'enjoyable', and I suspect that what he most enjoyed about it was winning.

HAL is vain to the point of petulance. Consider his soft-spoken 'outburst' when being interviewed at long distance by a BBC reporter:

BBC REPORTER: Good afternoon, HAL. How's everything going?

HAL: Good afternoon, Mr Amer. Everything is going extremely well.

BBC REPORTER: HAL, you have an enormous responsibility on this mission, perhaps the greatest responsibility of any single mission element. You're the brain and central nervous system of the ship, and your responsibilities include watching over the men in hibernation. Does this cause you any lack of confidence?

[*At this suggestion, the computer reacts as if stung. Behind his politeness, he seems to regard the question almost as an insult.*]

---

* Please see the Notes to this chapter at the end of the book for a description of this game.

HAL: Let me put it this way, Mr Amer. The 9000 series is the most reliable computer ever made. No 9000 series computer has ever made a mistake or distorted information. We are all, by any practical definition of the words, foolproof and incapable of error.*

Douglas Lenat, president of Cycorp, in Austin, Texas, is trying to build a machine with the intelligence of HAL. But he does not agree that it must have emotions, except in the sense of being able to decide between alternative courses of action. He believes that such a machine would need only 'wisdom and knowledge'.[8] He is horrified by HAL's statement to the BBC reporter that I quoted above:

> I certainly have never forgiven HAL. We all felt bad when he terminated the cryogenically slumbering crew, cut Frank adrift, and almost murdered Dave. But that's not what I found so unforgivable. It was not [his murders of the crewmen] that I found so unforgivable. To me HAL's biggest crimes were his conceit and his stupidity.
>
> By conceit, I mean claims like, 'No 9000 series computer has ever made a mistake.' This is more than just arrogant, more than just false; it is the *antithesis* of realism. If you met a man who genuinely believed he never had, nor ever would, make a mistake, you'd call him insane.† Surely NASA would never have trusted the mission to such a patently insane computer.[9]

As the year 2001 approaches, how near are we to producing a machine with the powers of HAL that is *not* 'patently insane' or prone to murder? There is certainly no hope of building such a machine by 2001, but by 2101 it should prove a very different story. Progress in computing proceeded at a staggering rate in the

---

* This remark of HAL's does not explain why his source code contains a routine (activated when he murders the hibernating astronauts) that says: 'Display the message: COMPUTER MALFUNCTION'! HAL's definition of the word 'error' remains mysterious.

† A psychiatrist might diagnose this condition as paranoia combined with violent homicidal tendencies. One is reminded of Hitler screaming at the Austrian Chancellor: 'I have never told a lie in my life!'

second half of the twentieth century, and shows no signs of slowing. Indeed, it is said that if progress in aviation had proceeded at anything like this rate, it would now be possible to cross the Atlantic supersonically for a fare of less than a cent.

There have been two great computer revolutions. The first, in the early forties, produced titanic devices like COLOSSUS, built in Britain to crack wartime German codes, which had about the same processing power as a modern laptop. But it, and its sister machines, were less easy to use than laptops! It was said of the ENIAC, the pioneering American machine built in 1944 at Princeton, New Jersey, to calculate artillery shell trajectories and design atomic weapons:

> It was too big. In fact it was worse than big, it was colossal, a veritable dinosaur of tubes and wiring. It was 100 feet long, 10 feet high and three feet deep. It had over 100,000 parts, including 18,000 vacuum switches, 1,500 relays, 70,000 resistors, 10,000 capacitors, and 6,000 toggle switches. There seemed to be no end to the thing, and [its designer John] von Neumann used to joke that just keeping it going was 'like fighting the Battle of the Bulge every day'.* When the machine once ran for five days without a single tube failing, the inventors were in hog heaven.[10]

The mightiest descendant of ENIAC (COLOSSUS had no descendants; its existence was kept secret for 30 years after the war so that it could be used against the Soviets) was the first supercomputer, ILLIAC 4, designed at the end of the sixties, about the time that Clarke and *2001*'s producer Stanley Kubrick were designing the fictitious HAL. It had a total memory storage capacity of a million bytes – or characters of memory – a fraction of the typically hundreds of millions of bytes in a modern personal computer.[11] It was almost immediately followed by the second great computer revolution, which was triggered by the Moon-landing missions.†

---

* The ENIAC consumed so much power that, according to legend, when it was switched on, it dimmed the lights all over town.

† As I explained in my book *The Next 500 Years*, the worldwide spread of personal computers and computing devices has so far proved the single most important legacy of the Moon landings. The Moon, with its labyrinthine craters

After a few years of false starts, the first personal computer, the Altair 8800 (named after the 12th brightest star), appeared in 1975. It was regarded by computer professionals as little more than a toy. Its programs were stored on tape cassettes, and even the tiny programs it was capable of handling took about 20 minutes to load compared with the fraction of a second that a modern computer takes to load a program more than 100 times bigger from floppy or compact disk. Its memory capacity was only 4,000 bytes, the length of a short newspaper article.

The growth in computing technology is accurately following Moore's Law, laid down by Gordon Moore of the computer company Intel in the sixties, that computer processing power doubles every 18 months. In 1998, computing power was thus approximately a billion times greater than it was in 1968, when HAL first appeared.[12] Computers can talk, but they do not yet understand what they are talking about. When I log on to the network America Online, my machine says: 'Welcome!', and when I log off, it says: 'Goodbye!' Some cars complain irritatingly that their owners have not yet fastened their seat belts; computer encyclopedias talk enthusiastically about their contents, and the newest computer games are filled with spoken dialogue. At Bell Labs in New Jersey there is a computerised face that 'shows emotion' and whose mouth emphasises different syllables as it talks.*

But either this speech is 'synthesised', so that words are generated by the machine itself to suit the occasion, or else they are inserted by actors and human programmers.[13] And machine speech still sounds scratchy and mechanistic, far from the warm, emotional tone of HAL (uttered by actor Douglas Rain). Devices

---

and cliffs, was a dangerous place to land on at high speeds. And so the astronauts, being 2½ seconds radio round trip time from Earth, had to have their own high-speed *on-board* computer to calculate their descent. And to fit into their cramped landing vehicle, it had to be far smaller than any computer yet built. Private industry copied this effort, hence the modern electronics industry.

* The first demonstration of machine speech, at a meeting of the French Academy of Sciences in 1878, nearly ended in violence. The physicist Du Moncel turned on a phonograph, a precursor of the gramophone, and it began to speak. Whereupon the celebrated 82-year-old physician Jean Bouilland leaped at Du Moncel, grabbing him by the throat. 'You wretch!' he shouted. 'How dare you try to deceive us with the ridiculous tricks of a ventriloquist!'

can be attached to word processors that make the machine utter the words that appear on the screen, and it is increasingly possible to dictate to them orally. But no machine can yet turn its own thoughts into words or understand a complicated thought beyond a simple command of one or two words. Nor will this be possible without the understanding of *context*. The sentence: 'Let's talk about how to recognise speech' can all too easily sound to a present-day computer like: 'Let's talk about how to wreck a nice beach.'[14] To be intelligent, it has to know, as a human does from the context of a conversation, whether the subject under discussion is likely to be speeches or beaches.

But all these problems are being steadily solved. There has been a massive switch of emphasis by those who are trying to make computers work more intelligently.* Computer scientists in past decades made the mistake – depending on one's point of view – of creating machine intelligence in specialised tasks, like playing chess, finding oil deposits, and diagnosing diseases. But the main effort now is towards improving the 'operating system', by which the computer organises its tasks, is able to do two or more different things at once – a technique known as 'parallel processing' – and communicates with its user in plain language. The widely used operating system Windows 95 is a promising example of this. Douglas Lenat sees three generalised steps towards building a HAL-like machine:

1. Program into the machine millions of everyday terms, concepts, facts and rules of thumb that comprise common sense.
2. On top of this base, construct the ability to communicate in a natural language such as English. Let the HAL-to-be use that ability to enlarge its knowledge base vastly.
3. Eventually, as it reaches the frontier of human knowledge, there will be no one left for it to talk to, so it will need to perform experiments to make further headway.[15]

Many computer intelligence specialists, academics in particular,

---

* How does one define 'intelligence' in a machine? The best definition I know of is that it is exhibiting intelligence when it does something that would *be considered intelligent* if done by a human being.

believe that a machine cannot truly think unless it has 'consciousness', awareness of the outside world. Professor Sir Roger Penrose, in his 1989 book *The Emperor's New Mind* – although exciting in his discussion of the *nature* of consciousness – seemed obsessed with this idea.[16] As Alan Turing suggested in 1950, an intelligent machine should even be capable of falling in love. This and similar ideas lie at the heart of the Turing Test, in which a human would have a prolonged conversation by keyboard with a computer in a separate room, and, if the machine was sufficiently 'intelligent', imagine that he was conversing with another human.[17]

But this approach today seems naïve. Surely all that a machine needs, to do things that would be considered intelligent if done by a human, is a *goal*, like the goal of a chess-playing machine whose sole aim is to checkmate the opposing King. The goal, of course, can be much broader than this, that for example of getting a starship crew safely to their planet of destination. The machine does not need the ability to converse in a civilised Oxford-high-table fashion on any academic topic that is raised. It only needs the ability to foresee dangers and solve problems.[18]

It may be that super-intelligent machines of the future will use a different kind of logic from that used by the computers we are familiar with. Their chips, instead of being *digital*, as are today's, will be *analogue*. The difference is that instead of processing data rigidly, as a digital computer does, it will draw inferences and reach conclusions from incomplete data. A digital computer always does the same thing – what it has been programmed to by humans – whenever it is switched on, but an analogue computer could in theory *write its own programs*.

It would do this by 'genetic algorithms', lines of code in the program which in effect 'mate' and have 'offspring'. The new line of code, created by the computer itself and not by the human programmer, then influences the machine's behaviour in ways that the programmer might not have foreseen. With a digital computer, the programmer tells a machine precisely how to solve a problem, but an analogue machine is expected to solve a problem without being told how to do it. It is the difference between strict conformity and intellectual freedom. 'Digital design is like building with Lego blocks,' said Richard Chesson, an engineer at ST Microelectronics of Milan. 'Analogue design is more like wood carving.'[19]

Whether analogue or not, more independence and intelligence could come from a computer if its operating system could be made more flexible. One example of how such flexibility might work is taken from theology. It comes from the Doctrine of the Holy Trinity, as espoused in the fourth century AD by Saint Athanasius (and afterwards condemned as a heresy because it was seen as challenging the authority of God). In this doctrine, the three personages of the Trinity, the Father, Son, and Holy Ghost, are 'co-equal'. None of the three has greater or less authority than the other two.

Athanasius did not mention the possibility that the three might have disagreements, or what happened when they did. But William McLaughlin, an astrophysicist at the Jet Propulsion Laboratory at Pasadena, California, does. He sees the concept of the Holy Trinity as a 'revolutionary form of logic which is ideal for new forms of computer systems'.[20]

A digital computer, whether a mainframe or the kind we have on our desks, has a single central processing chip. It is a hierarchical system, with all authority emanating from this entity. But in McLaughlin's proposed system based on the Holy Trinity, there would not be one but *three* central processing chips. When they were in disagreement they would take a majority vote. They would work out new solutions to problems by themselves. They might be able to do things without being told by a human how to do them.

Nevertheless, even when such miraculous-seeming machines are built, it will be their morals, or their possible propensity to misbehave disastrously, that will need the closest attention if we are to believe them when they say: 'I enjoy working with people.' This danger was recognised back in 1893 when Ambrose Bierce wrote a short story about a chess-playing 'automaton' that behaved like HAL at his worst:

> Presently [its inventor] Moxon, whose play it was, raised his hand high above the board, pounced upon one of his pieces like a sparrow-hawk and with the exclamation 'Checkmate!' rose quickly to his feet and stepped behind his chair. The automaton sat motionless.
>
> I then became conscious of a low humming or buzzing which grew momentarily louder and now more distinct. It seemed to come from the body of the automaton, and was

> unmistakably a whirring of wheels. Before I had time for much conjecture as to its nature, my attention was drawn by the strange motions of the automaton itself. A slight but continuous convulsion appeared to have possession of it. Its body and head shook like a man with palsy until the entire figure was in violent agitation.
>
> Suddenly it sprang to its feet, and with a movement almost too quick for the human eye to follow shot forward across the table and chair with both arms thrust forward to their full length – the position and lunge of a diver. Moxon tried to throw himself backward out of reach, but he was too late: I saw the horrible thing's hand close upon his throat. Moxon was now underneath it, his head forced backward, his eyes protruding, his mouth wide open and his tongue thrust out; and – horrible contrast! – upon the painted face of his assassin an expression of tranquil and profound thought, as in the solution of a problem in chess.[21]

Or, to conclude on a more positive note, perhaps the only thing we shall have to worry about will be HAL's memory. Knowing the habits of engineers, it is hard to believe that he became operational on 12 January 1997. That day was a Sunday.

Chapter 15

When Luxury is Necessity

> How many times when you are working on something frustratingly tiny, like your wife's wrist watch, have you said to yourself: 'If only I could train an ant to do this!'
>
> Richard Feynman, 'There's Plenty of Room at the Bottom'

One of the least agreeable of experiences is a long flight in a crowded plane, from one side of the world to the other, especially when travelling economy class.

The food is unwholesome, the air is bad, with low oxygen content – the airline may be cutting it to save fuel – and there is a risk that being unable to move for so long will cause potentially dangerous blood clots. There is also the risk of impurities in the air, highly concentrated in the most crowded parts of an aircraft – of germs, mould, yeast, acetone, isopropyl alcohol, and compounds found in fuels, cleaners, plastic and solvents of grease.[1] At the end of the flight, which may have lasted up to 16 hours, one almost always feels despondent, tired, irritable, and unfit to do useful work or make important decisions.*

Many people who are sceptical about interstellar travel, even to

---

* The American Secretary of State John Foster Dulles admitted making a catastrophic decision in 1956 when tired and irritable after getting off a long flight. Learning that the Egyptian dictator Nasser had made an arms deal with the Soviet Union, he immediately cancelled an American loan to finance the Aswan High Dam in Upper Egypt. This precipitated the Suez Crisis of that year in which the affairs of several nations were convulsed.

the nearest star, slightly more than four light years away, ask how it would be possible to endure such conditions not for 16 hours, but for more than *two years*.

It appears inconceivable that anyone would willingly do so unless they were accorded not merely the comforts of first-class airline travel, but at least those of first class on an old-time ocean liner.

Except for passengers in cruise liners, very few people today have experience of these luxurious vessels which, until the mid-fifties, were the normal way to make long journeys. One could walk around, breathe fresh air, take part in dances and banquets, and walking half a dozen times round the deck made an invigorating one-mile stroll. Here is one splendid description of such a ship, the German 54,000-ton liner *Vaterland*, in its time the world's most luxurious ship. It was built to carry up to 3,909 passengers and 1,245 staff to look after them. It made its maiden voyage in 1914 from Hamburg to New York, where it attracted glowing descriptions on arrival:

> It ranked as a ship of exceptional luxury, a veritable floating palace. In step with the latest technology, Frahm anti-rolling devices kept the ship on even keel and special insulation virtually eliminated engine vibrations, allowing passengers relaxed comfort as they danced in spacious ballrooms or exercised in the Roman-style marble swimming pool or dined at the Ritz-Carlton Restaurant.
>
> Unique among modern ships, the funnels did not pass through midship but were bifurcated below passenger decks joining at the top, thus allowing large spacious rooms in midship. An ornate Mewés-designed skylight graced the Wintergarten Social Room which featured rich Persian carpeting, heavily upholstered divans and square window casings identical to a luxurious hotel. The ship's kitchen staff of a hundred cooks and bakers had at their disposal an array of electric equipment, including electric potato peelers and dishwashers. Four hundred stewards eagerly provided every service. Elevators whisked passengers to its many floors, and fifteen thousand electric lights provided illumination for the staterooms and public rooms, among them the huge first-class dining room that seated eight hundred guests.[2]

For purposes of comparison with a starship, we can dispense with the 100 cooks and bakers and the 400 stewards, although the future equivalents of electric potato peelers and dishwashers will certainly be needed. For the most part, the jobs done by such people will be performed by robots. It is the sheer physical and psychological necessity of spaciousness that must be emphasised. Whether the vessel should be spacious will not just be an argument between extravagance and economy. These apparent luxuries will be necessary so that the colonists do not arrive on an alien planet unfit to do any useful work.

There is a definite practical reason why spaciousness will be needed. The enemy of good health in space is weightlessness, or rather the one millionth of terrestrial gravity that astronauts experience in Earth orbit. We have seen how Russian cosmonauts, after spending more than a year in space, are unable to walk when they land. The most serious effect is the loss of calcium in their bones which may take months to return to normal. To counteract this, it will be necessary in a starship – as in all long-duration manned space voyages – to have artificial gravity.*

Like the giant ring-shaped space station in the film *2001: A Space Odyssey*, the entire region of the crew quarters will be made to rotate so that centrifugal force will create gravity in its outer regions. But this will only work if the rotating parts of the crew quarters are at least 160 metres wide.[3] Otherwise the rotation will cause giddiness and sickness through the so-called Coriolis force.† Spaciousness and comfort are thus forced upon the crew by necessity. Since the crew quarters *must* be at least 160 metres wide, there will be plenty of room in them for all the luxurious amenities that people could desire during a voyage that lasts several years. The space available to them will be comparable to that of the passengers in the *Vaterland*. Since, as I explain

---

* Einstein would surely have frowned at the expression 'artificial' gravity. It would not have been clear to him what was artificial about it. If it *appears* to be identical to gravity, he would have said, then it *is* gravity. The fact that it has been artificially *created* would have seemed to him irrelevant.

† From the calculations of the French mathematician Gaspard Coriolis (1792–1843). There is a phenomenon first predicted by Coriolis that, because of the Earth's rotation, bath water forms opposite spirals as it runs into the plug hole according to whether one is in the northern or southern hemisphere. This actually happens, but only on a tiny scale. Being overwhelmed by other effects, it is never observed.

below, at least half the crew will be asleep at any one time, there will be space for a recreational park, with plants and trees for aesthetic reasons and to freshen the air, that will never be too overcrowded.

How will people eat? On an airliner, all food is stored on board to be served at the appropriate meal times. But this would not do in a starship. Suppose, as I have said earlier, that there are 900 to 1,000 colonists on board. If each of them requires two kilograms of food per day – not to mention water – then, for a two-year voyage, they would need more than 1,000 tons of stored food and provisions, adding significantly to the mass of the ship and the energy needed to propel it. And when we add to this consideration the need for storage space, refrigeration and intelligent retrieval (people will not want the same dish every day), the cost becomes insupportable.

Instead, it is likely that food will be manufactured in a device that I will call an *atomic food processor*; and it will be manufactured atom by atom from the ship's refuse.

Before examining this extraordinary idea, it is necessary to look at the fundamental nature of substances. All matter that exists on Earth can be divided into two classes. In one class are such things as air, water, metal, glass, concrete, and sand. This is *inorganic* matter. It is lifeless. But the other class of substances is made from plants or living creatures: olive oil, petroleum, sugar, starch, glue, gelatin, silk, coal, rubber, paper, and penicillin. This is *organic* matter, all of which is made from compounds of carbon.[4]

The discovery of this distinction by the Swedish chemist Jöns Berzelius in 1807 marked the beginning of modern chemistry, of which 'inorganic' and 'organic' are two separate branches. Most people are only familiar with two compounds of carbon: carbon dioxide and carbon monoxide. But in fact there are more organic compounds – compounds containing atoms of carbon – than there are of all the other elements put together. No one knows for certain, but the number of *possible* carbon compounds may be infinite.* As Isaac Asimov explains in his book *The World of Carbon*:

---

* The pioneering 60-volume work *Handbuch de organischen Chemie* ('Handbook of Organic Chemistry') by the German chemist Friedrich Beilstein (1838–1906) listed and described more than 1,500 carbon compounds in 11,000 pages. For

> There are many more compounds that contain carbon than compounds that do not. Why is this? Well, when other atoms (other than carbon) are hooked together to form molecules, the best results are obtained when only a few atoms are involved. A molecule made up of only two or three atoms is often quite strong and the atoms hold together firmly. As more atoms are added, however, the molecule becomes rickety and it is more and more likely to fall apart. An inorganic molecule containing more than a dozen atoms is, for that reason, quite rare.
>
> Molecules containing carbon atoms are an exception to this general rule. Carbon atoms can join one another to form long chains or numerous rings and then join with other atoms as well. Very large molecules may be formed in this way without becoming too rickety to exist. It is not unusual for an organic molecule to contain a million atoms.[5]

What does this all have to do with food? It turns out that four out of the five main requirements of nutrition – proteins, fats, carbohydrates, and vitamins – are made from compounds of carbon. (Of vitamins more in a moment.) Even the fifth, minerals, depend on carbon for their existence in edible form. Calcium, for example, comes mainly from milk and cheese, whose chemical structure is based on lactic acid, a compound of carbon, hydrogen, and oxygen. In turn, iodine comes from seafood, and iron from liver and green vegetables. But carbon compounds are not all. The atoms of other elements that are essential to a balanced diet can also be manipulated in the food processor. This is especially important for minerals. Iodine, for example, which makes the thyroid gland function normally, is found in bananas, cheese, cherries, and milk. Potassium keeps bodily fluids stable and is present in significant amounts in natural foods. Iron, which prevents anaemia and transports oxygen to the tissues, is found in aubergines, beans, and other vegetables, as well as such fruits as prunes and raspberries. And phosphorus plays an essential role in all body cells.[6] But I am glad not to have the

---

most of this century no equivalent work existed in English, and organic chemists had to be fluent in German. Today, more than seven million such compounds are known and are classified in English in electronic databases.

responsibility for programming the computer that will manipulate the ship's refuse to produce the atoms and molecules that will create all these foods!

How will it be done? In 1959 Richard Feynman, considered by some the twentieth century's second greatest physicist after Einstein, gave a lecture, 'There's Plenty of Room at the Bottom', proposing an entirely new method of constructing matter from unimaginably small raw materials. It would be made atom by atom, or molecule by molecule, instead of the usual way we make things, based only on the building blocks we can *see*. As he put it:

> The principles of physics, as far as I can see, do not speak against the possibility of manoeuvring things atom by atom. It is interesting that it would be, in principle, possible for a physicist to synthesize any chemical substance that the chemist writes down. Give the orders, and the physicist synthesizes it. How? Put the atoms where the chemist says, and so you make the substance.[7]

Since then, enormous progress in 'nanotechnology' – from the word 'nano', meaning a billionth – has been made. (Indeed, the very word is a gross understatement. It should perhaps be 'tenth of a trillionth technology' – a tenth of a trillionth being a decimal point followed by 12 noughts and a 1 – roughly the diameter in centimetres of an atom.)* In 1981, Gerd Binnig and Heinrich Rohrer of the IBM Zurich Research Laboratory invented the scanning tunnelling microscope, which allowed the imaging of individual atoms, an achievement that later won them the Nobel prize in physics. And four years later, Binnig and Christoph Gerber, also of IBM in Zurich, along with Calvin Quate of Stanford University, invented the atomic force microscope. This has allowed not only the direct imaging and manipulation of living cells, but also the construction of gear systems and motors that are made of only a few thousand atoms, no mean feat

---

* Nanotechnology will have its sinister side. It could be used to create deadly weapons against which there would be no defence. As Ed Regis says in his book *Nano*, invaders the size of bacteria could be sent over a national border on a gust of wind, enter people's bodies, and turn them into slime.

considering that there are more atoms in a teaspoonful of water than there are teaspoonfuls of water in the Atlantic Ocean.[8]

For 2,000 years, from the proposal of the Greek philosopher Democritus that all matter was made of tiny, invisible things called atoms (from the Greek word 'indivisible') to the beginning of the twentieth century, it was believed that atoms did not really exist, that they were merely a convenient mathematical abstraction. 'Atoms and molecules are a mystical conception,' said the French chemist Pierre Berthelot in 1877. 'Who has ever seen a gas molecule or an atom?'[9] But by 1989, progress had enabled K. Eric Drexler, the great pioneer of nanotechnology, to speculate about 'molecular manufacturing, entailing the ability to build molecular systems with atom-by-atom precision'.[10]

And so to the atomic food processor manipulating carbon compounds that will feed the crew of a starship without the necessity of storing food. As Ed Regis says in his 1997 book *Nano*, extrapolating from an idea of Drexler's:

> You could invent this black box, a 'meat machine' or something of the sort that could physically transform common materials into fresh beef. The machine might be about the size and shape of a microwave oven, for example, and it would work the way a microwave oven does, more or less. You'd open the door, shovel in a quantity of grass clippings or tree leaves or old bicycle tyres or whatever, and then you'd close the doors, fiddle with the controls, and sit back to await results. Two hours later, out would roll a wad of fresh beef.[11]

But more than beef and other proteins will be needed for good health. One essential output of the atomic food processor will be vitamin C, found in fruit, vegetable, and salad, which prevents the debilitating disease of scurvy.

Countless seamen on long voyages suffered or died from scurvy until Captain Cook, in his 1769 voyage to Australia, set an example by giving his men cress and orange juice.* The symptoms

---

* The threat of scurvy on the seas finally ended in 1795, when the British Navy required all their crews to drink lime juice, giving Britons the lasting nickname 'limeys'.

of scurvy, although easily cured, are fearsome. As the sixteenth-century medical writer Philip Stubbes complained:

> Do we not see by experience that those that give themselves to sweet meats are never in health? Doth not their sight wax dim, their ears become hard of hearing, their teeth rot and fall out? Doth not their breath stink, their stomach belch forth filthy humours and their memory decay? Doth not the breathing become laboured and the body corpulent, yea, sometimes decrepit therewith, and full of all filthy corruption?[12]

The last thing a starship commander will want is for his crew to land on another planet in this condition.

When I first started thinking about manned starships, I imagined gloomily that a crew of hundreds on a voyage lasting years would need great herds of cattle, not to mention fields of agriculture for growing vegetables, and orchards for fruit. (They could not live off the resources of space, since out in interstellar space – as opposed to within the solar system – there will be no resources to live off. The ship will be absolutely alone. No voyagers in the loneliest seas will ever have been so cut off from the rest of humanity. Once the solar system has been left far behind, the chances of passing near a planet other than the planet of destination, or an asteroid filled with useful organic compounds, are virtually zero.) The combined weight of these animals and plants, and the land needed to accommodate them, would have added tens of millions of tons to the mass of a ship. Then came the realisation that all food could be pre-cooked, frozen, and preserved, as in an airliner, which would have brought this extra mass down to a few hundred tons. But now, with atomic food processors, even this can be dispensed with. The weight of the crew's food technology can be brought down to virtually nothing. When the ship's refuse, anything that is made from organic compounds, from old sacking to bodily wastes, can be made into edible food, the only weight required for nutritional requirements will be the food processing machines themselves and the computer needed to operate them.*

---

* This may all seem wildly optimistic, and I do not mean to imply that anything like it can be done today. Nobody at present has the faintest idea how to

But still we have the prospect of 900 people walking about, with all the additional weight their activities will create. It will perhaps increase the comforts of those on board if half of them are in deep sleep at any given time.

It has often been suggested by writers contemplating extremely long journeys in space that the crew should be deep-frozen and put into a coma-like sleep for the duration of the voyage. They would become what some science fiction writers have called 'corpsicles'.[13] They would hibernate and be awakened when arrival was imminent. This, after all, would save a great deal of space and expense. It would not be ordinary sleep. Their bodies would be frozen so that all ageing activity ceased. But experiments with rats have shown that this procedure may be inherently dangerous. And in a starship, even a momentary loss of electric power, as is always liable to occur, could kill the sleepers. Moreover there is a risk, as in the novel *Beowulf's Children* by Jerry Pournelle, Larry Niven, and Steven Barnes, that the atoms in their brains would be in the wrong places when they were awakened, and that they would never again be normal people.

It is not that the technology to put people into deep sleep for a period of years is unlikely to be developed in coming centuries. But it appears more appropriate for medical experiments on Earth than for starship voyages. It is a question of reasonable risk.

There are several research projects to enable people with currently incurable diseases to be put to sleep so that they can be awakened at some future time when a cure is perfected. But those who submit themselves to such experiments will be risking very little. If they *don't* do it, they will die of the disease. And if they *do*, there is a reasonable chance of successful awakening and cure. But in a starship, the odds are very different. By allowing themselves to be put into a deep-frozen sleep, the colonists would be risking *everything*. If there is any technical failure, and a substantial number of them either cannot be awakened or can only be awakened as zombies, then the entire voyage becomes a

---

manipulate atoms to the extent of turning old sacking into barbecued chicken, or urine into wine! All we have so far is the confidence that doing so would not violate any physical laws, together with several remarkable achievements in micro-micro-miniaturisation. Nevertheless, it is a principle of physics that anything is possible that is not forbidden by fundamental laws.

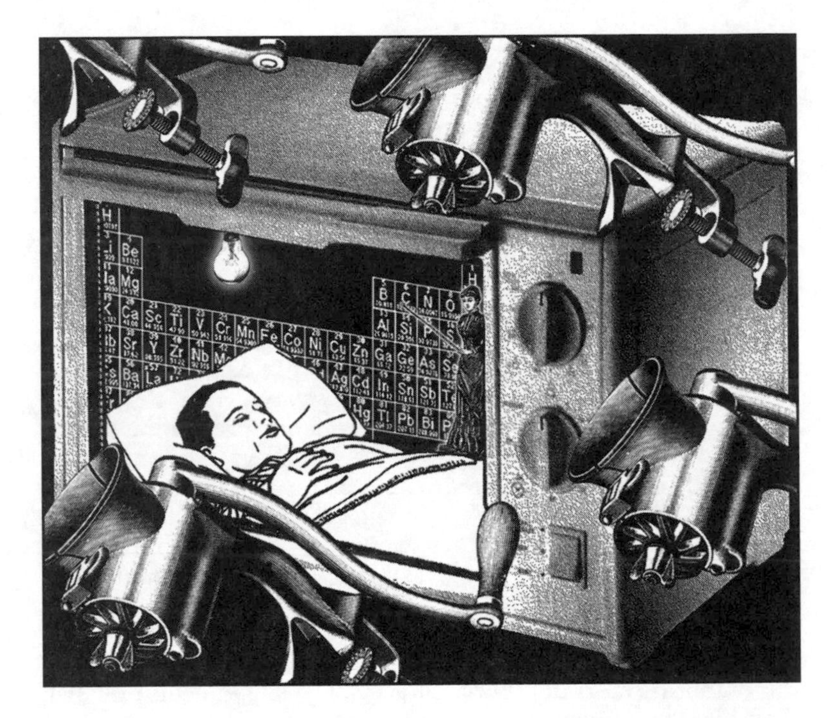

catastrophic failure. (But this technique may be desirable in extremely long flights: see Chapter 17, 'The Big Sleep'.)

A lighter form of sleep will be much more suitable. It will be convenient to put the colonists under a mild anaesthetic for periods of months at a time. To save air, food, living space, and water, perhaps half the crew will be asleep, but not deep-frozen with all the complex and possibly dangerous technology that would entail. Sleepers will have a tube down to their stomachs providing water, food, and the sleeping drug. All the sleepers will be connected to tubes that will take care of bodily wastes.[14]

And what of those who are awake? There is an ailment on Earth called SAD, short for 'seasonal affective disorder', from which a tenth of the population is said to suffer and which is caused by lack of sunlight in winter. Its general symptoms are lethargy, depression, insomnia, and a proneness to more serious illnesses.

The bodies of people who live most of their lives under the yellow, artificial light that is used in homes and offices cannot manufacture vitamin D, which both inhibits these afflictions and strengthens bones, prevents tooth decay, and produces the chemical serotonin that strengthens the immune system and makes people feel happy and self-confident.*

Travellers in a starship could all too easily fall into this condition. There will be no 'natural' light, since once the ship has passed beyond the outermost limits of the solar system, the brightness of the Sun itself will have dwindled to that of a star of medium brightness.† Outside the ship, except for the comparatively feeble light of the stars, there will be limitless darkness. And as the ship gradually accelerates towards its cruising speed of 92 per cent of the speed of light, even the stars themselves will vanish (see Chapter 12, 'The Star on the Starboard Beam').

The remedy for this, for those who are not asleep, will be *artificial* sunlight. For 12 hours each day, the period of terrestrial daylight in the tropics, the ship's interior will be made to look like a good stage production of Oscar Wilde's play *The Importance of Being Earnest*, whose third act takes place in a garden on a summer afternoon. This will be made possible by metal halide lights, a type of fluorescent lighting whose filament is so hot that its radiation is entirely white. Like the Sun's radiation, it encompasses the entire spectrum from the ultraviolet to the infrared. There is no actual proof that metal halide lighting manufactures vitamin D in animals, but it certainly behaves as if it does. I have seen it nourishing fish and amphibians in the marine section of London's Kew Gardens. There, underground, with the metal halide lights shining overhead for 12 hours each day, these animals enjoy the health and lifespans that they would have in tropical pools and beaches.

---

* In the Norwegian town of Tromso, well inside the Arctic Circle, people celebrate 'Sun Day' in the spring, after enduring 49 sunless days, with prayers, a public holiday, and other displays of emotion. And in Afghanistan, the fanatical Islamic government known as the Taleban has been accused of genocide against women by forcibly keeping them indoors and even depriving them of sunlight by ordering their windows to be painted black.

† On Pluto, which in 1999 swung out beyond Neptune to become the Sun's most distant planet, 40 times further from the Sun than the Earth is, the Sun cannot even be seen as a disc. It appears only as a bright star that throws a faint and melancholy light on the icy surface.

There is every likelihood that humans will enjoy the same sense of well-being under its influence. It is interesting that metal halide lighting is used to illumine the interiors of nuclear submarines, which spend many months submerged, and where ordinary domestic lighting would be intolerably dreary. (The only reason why it is not used in homes is that it is about three times more expensive than ordinary lights.) The journalist Rachel Kelly, who spent an hour under such lamps during a dreary November in London, reported: 'It was an utterly hedonistic hour, and I felt serene and relaxed.'[15]

This lighting will be everywhere in the crew quarters. I have mentioned how, in the ocean liners of old, people exercised by walking several times round the deck. There will surely be corridors in a starship in which those who are awake can do the same. (Even on aircraft, during long flights, passengers are encouraged to get up and walk around. The drawback is that in economy class, if it is crowded, it is almost impossible to do this.) There will probably, in addition, be a 'holodeck', as depicted in *Star Trek*, a region created by virtual reality which does not in fact exist and which changes according to the wishes of each person, who can act out their fantasies electronically.[16] (See the next chapter.) But you cannot take *physical* exercise in a holodeck since nothing in it is physically real. It is no substitute for corridors and public rooms, and for the spaciousness that these imply.

Reducing comforts to save money will not be a sensible option for starship designers. What seems to be luxury will truly be necessity. Confinement for two years or more in a metal capsule might seem a grim prospect. But with natural lighting, spacious quarters, Earthlike gravity, and – if the atomic food processor's computer is well programmed – limitless food and drink of high quality, there is a good chance that the colonists will reach their destination world neither ill nor dead nor insane. There, as I explained in Chapter 1, their standard of living will deteriorate abruptly. The moment they set foot on another planet and attempt to make it their home, they will be transported in an instant from the living standards of the twenty-second or twenty-third century to those of the nineteenth. It may be that they would lack the strength to endure the latter without having been fortified by the comforts and amenities of the former.

Chapter 16
The Game's Afoot, Watson

The floating arrow of light that had been their mysterious guide through the labyrinths of the Crystal Mountain still beckoned them on. They had no choice but to follow it, though, as it had done so often before, it might lead them into yet more frightful dangers.

Arthur C. Clarke, *The City and the Stars*

Drama is real life with the dull parts left out.

Alfred Hitchcock

The characters in *Star Trek* constantly have virtual 'adventures', elaborate computerised games, to relieve their boredom. On this point, one cannot fault the series. There will have to be highly original 'inflight entertainment' on board a real starship; or else the crew and passengers, on voyages lasting *years*, will become very stressed. Enforced idleness demands intellectual stimulation, and the more intelligent people are – and the people on board a starship will be highly intelligent – the more this will be true.

The nearest analogy today to being on a starship is spending a winter at the South Pole. It is a close enough parallel. Instead of being surrounded by the boundless vastness of interstellar space, people are in the midst of an empty frozen wilderness that stretches for hundreds of kilometres in all directions. To keep themselves sane, the scientists at the Amundsen–Scott Station enjoy amenities which some might consider extravagant luxuries, since the station is state-funded, but which they regard as essentials. These include a bar, a library of both books and videos, a gymnasium, a lounge with a billiard table, and, in the words of an engineer who has wintered at the Pole, a 'raucous derivation of volleyball. In some ways the environment is comparable to outer space, but you amaze yourself how well you acclimatise yourself to functioning there.'[1]

But although people on a starship will no doubt have all these things, they will have the technology to entertain themselves in a far more sophisticated manner: they will pass much of their time playing computer games.

This is no frivolous suggestion. Some of the most important people attached to the Russian and American manned space programmes are psychologists, who strive to combat boredom. In space, one is just as lonely as at the South Pole. 'You can't run out and get a pizza or watch your favourite TV programme,' said American astronaut Norman Thagard, who spent four months in the Russian *Mir* space station in 1995.[2] NASA employs six such expert advisers and the Russians no fewer than 30.[3] After a mere two months in space, Russian cosmonauts – who have been spending much longer up there than the Americans in their space shuttles – tend to become dispirited and lethargic. To counter these symptoms, they take up books, fruits, magazines, and delicacies, none of which are needed for the technical achievements of their missions, but which are essential to good morale.[4]

Bad morale can be very dangerous on a flight lasting years. It would not remain static; it would continue to deteriorate. It could lead in time to mutiny or even insanity. A good example of this in science fiction was given in Robert A. Heinlein's 1963 novel *Orphans of the Sky*, in which the crew of a starship have forgotten that they are even on a ship and descend to a state of rigidly stratified and superstitious social organisation.[5]

NASA has always taken the problem of boredom on long space flights very seriously. During a classic experiment in 1970 I saw for myself that their concern was justified. The space agency co-sponsored an experiment called Project Tektite, in which teams of scientists lived in turn for weeks at a stretch in a submarine habitat fixed to the seabed off the American Virgin Islands. Despite the luxurious comfort of the habitat, and the demands of their work, which consisted of collecting and classifying marine specimens, they soon became quarrelsome. They quarrelled about the use of the 'wet room', a lobby leading from the entrance to the two other rooms. Two chemists in the experiment insisted on using it for storing their own scientific equipment and refused to allow their colleagues to do so. 'Our expensive underwater camera lenses were just pushed aside,' one of the latter told me furiously. 'Although we kept ourselves under control for the two weeks we were down there, I doubt whether we could have done so for a period of many months.' The trouble was that, apart from their work and telephone conversations with people at the surface, they had nothing with which to occupy themselves.[6] It was an interesting conclusion, and one that the space-travel-oriented designers of the experiment had in mind.

In a starship, there won't be anyone to talk to outside the ship. There won't *be* any 'surface'. The solace of talking to Mission Control and school pupils enjoyed by shuttle astronaut crews will be denied to the crew and passengers. Being light years from Earth, they cannot phone home or surf the Internet. They will be more alone than anyone in history. If they are not to go mad like Heinlein's characters or dangerously belligerent like the Tektite scientists, they will need leisure time, and *something*, apart from whatever work they have to do, that will absorb them utterly during that leisure time. Books, magazines and films – although in electronic form they will exist on board in plenty – cannot do this. They are too *transient*. They are absorbed too quickly. Over very long periods they will be insufficiently engrossing. Even a really long and exciting book like *The Decline and Fall of the Roman Empire* or *War and Peace* can be read in a few weeks and almost learned by heart in a few months. Something richer and more subtle will be needed.

Computer games will fulfil this need. Even if not true today, it will be true by the time that interstellar travel becomes feasible, and probably long before, that the best computer games will be far more intellectually satisfying than any book, magazine, or film could possibly be. For one thing, with countless situations to explore within the game, they will take weeks or months to play. They are like a film in which the player becomes one of the characters, thereby altering the story in an unlimited number of ways. I am not talking about games like solitaire, or naval battles, or any of the simple screen games that people play surreptitiously in offices, but full, interactive adventure games which, like chess, always end differently, and whose scenes are so realistic that they are indistinguishable from real life. The quotation from Arthur C. Clarke's novel *The City and the Stars* which began this chapter is part of a superb description of such a game.

In the past two decades, electronic games have undergone staggering technical progress. In the days before a personal computer could handle a mouse, digitised voices, music, or colour films, such games used to be distinctly dull. There were no pictures, and all action was initiated by typing out instructions and then pressing the Return key. A typical exchange might be:

> YOU ARE THE DICTATOR OF BARATARIA, AND A HOSTILE MOB IS APPROACHING YOUR PALACE. DO YOU . . .
>
> 1. Flee to Switzerland with the national funds?
> 2. Make a speech from the balcony?
> 3. Invite in a deputation?
> 4. Mow them down with machine guns?
>
> PRESS 1, 2, 3 OR 4 on the keyboard.

After a few more questions, the screen would clear and the message would appear:

> YOU'RE USELESS! THE COMMUNISTS HAVE SEIZED BARATARIA. ANOTHER GAME? (Y/N)

In 1979, millions of people played *Zork*, which was then the top fantasy adventure game. As the game began, the words flowed across the screen: 'You are standing in an open field west of a white house, with a boarded front door. There is a small mailbox here.' You were then invited to describe on the keyboard *in words* what you thought the invisible hero should do when placed in this situation.[7]

Today, things are very different. You barely need to touch the keyboard at all. Almost everything is done with the mouse, that most convenient instrument without which no modern personal computer is complete. Aiming and clicking the mouse is to stretch out one's hand to influence events in another world. Instead of boringly *verbal* descriptions of situations, the player sees moving pictures or often real film with real actors, which are sent off to perform different tasks at a click of the mouse. A turn-of-the-century computer game thus fulfils that age-old dream of watching a film and becoming one of the characters.

One of the latest games is as far removed from the earlier verbal games as bridge is from snap. Called *Sherlock Holmes and the Case of the Rose Tattoo*, it takes the player through an adventure in Victorian England which is as rich and complex as any of Sir Arthur Conan Doyle's stories.[8] All the player needs, in addition to a personal computer, is a mouse, a sound blaster, a CD-ROM drive, and a great deal of patience and cunning, and the game's afoot.

You become Holmes, moving him around with the mouse through dozens of elaborate locations, and you follow countless false trails while seeking the correct one.

The game begins in a realistically furnished 221b Baker Street. Mrs Hudson brings an urgent summons from your brother Mycroft to meet him at the Diogenes Club. As you and Watson arrive, the club is blown up and Mycroft is gravely injured.

'A gas explosion,' say the police. 'Gas explosions are the price of progress,' adds the insufferably unimaginative Inspector Lestrade. But Watson finds traces of a bomb in the club – after, with great difficulty, getting past the obstinately stupid porter guarding the ruins – and Holmes is soon on the trail of the would-be murderers.

In about 20 hours of play, the pair unearth more and more evidence of an international criminal plot, being led, as in a treasure hunt, from one clue to the next. Until they solve the current riddle, they cannot proceed.

Until you discover how to win his attention, a journalist keeps repeating: 'Some cretin in editorial has spiked my story!' Numerous objects must be analysed at Holmes's lab in Baker Street, always requiring different chemicals. There is a dragon of a matron to be appeased at Bart's Hospital who rants about 'discipline' and 'regulations' until you want to throttle her. And the high point of the game is a staggeringly difficult darts game in which Holmes must aim his darts by clicking the space bar at the right moment.

The Rose Tattoo is one of those games that, because they involve a visible 'hero', are known in the trade as 'interactive adventures'. They are almost as expensive to make as a B-movie. An ancient 'you-are-standing-in-an-open-field' game could be created by a single amateur programmer on his kitchen table, and many of them were. But with a modern interactive adventure, when you have paid all the set and costume designers, actors, researchers, musicians, programmers, accountants, de-buggers, and logic-checkers, you could be looking at a bill for several million dollars.

Interactive adventures have little in common with arcade games, which entail only quickness of the eye and hand as one steers a racing car or shoots down alien spaceships. The former, by contrast, demand imagination and intense thought as one moves from one mystery to the next.[9] As one creator of games explained:

> What makes the game hard or simple, or fun or boring is the level of anxiety – 'burning questions' you can keep bombarding the player with. To me, the question shouldn't be: 'How do I open the cabinet?' but: 'Will opening that cabinet solve this enigma for me?'[10]

And they are surely the wave of the future. The best American rival to *The Rose Tattoo* is *Gabriel Knight 2: The Beast Within*, a supernatural thriller in which two young detectives (one of whom is called Gabriel Knight) must track down a werewolf who is terrorising Bavaria.[11]

Both games have the merit of being highly educational. In the latter we attend the first night in Bayreuth of a lost Wagner opera – specially written and composed for the game – and we get a guided tour of mad King Ludwig's castle of Neuschwanstein. And the puzzles they contain often depart from what would be probable in real life to become ingenious and funny.*

Remember also Clarke's 'labyrinths of the Crystal Mountain',

---

* In one scene in the werewolf game, the hero must extract the keys belonging to the distrustful hall porter of a club so that he can explore its locked basement. He buys a cuckoo clock and sets it ticking at the back of the club so that the porter thinks someone is knocking at the back door. While the porter is absent

the floating arrow of light beckoning them on, that might 'lead them into yet more frightful dangers'. Among many computer game players there is a passion for three-dimensional labyrinths and mazes where moving arrows suggest alternative choices of direction. Considering that Clarke's story was written in the 1950s, long before the most primitive computer games existed, it showed extraordinary prescience. To navigate a difficult and perilous labyrinth in real life might be wearisome, but there is something deeply satisfying in doing it virtually. It is no coincidence that many of the best adventure games have these sinister

investigating, the hero extracts the keys. What I found hilarious was the idea of someone walking along the street carrying a two-kilogram cuckoo clock in his back pocket.

journeys. It does not matter whether the labyrinth is above or below ground. *The Beast Within* has its dark, wolf-infested forest; *The X-Files Game* has its tortuous warehouse with a gunman behind almost every doorway, and *Black Dahlia*, the story of a search for a medieval talisman, has a network of tunnels below a ruined monastery which the player must navigate by trial and error until he has visited all its hidden chambers without being slain by booby traps.

To understand the kind of games that will be available a century or so hence, one must appreciate the reasons for the tremendous improvement of the games described above compared with the much more primitive ones that preceded them.

It is mainly a question of the computer's memory. The greater the memory, the more complex the game can be. The first home computers, back in the 1970s, had memories up to 32,000 characters, or 'bytes', the equivalent of 5,000 words – and some of them had much less. Thirty-two thousand was merely the length of a long magazine article, and it gave no opportunity for character-driven stories or any sort of realism. A few simple pictures, considered revolutionary at the time, were created by keyboard characters. One could depict a battleship, for example, with straight lines by repeatedly pressing the shift of the hyphen and other keys. It would look something like the drawing below.

This may not look like a very workmanlike battleship (its gun firing shells towards the right of the screen in the form of asterisks), but it is the best I can do using my word processor

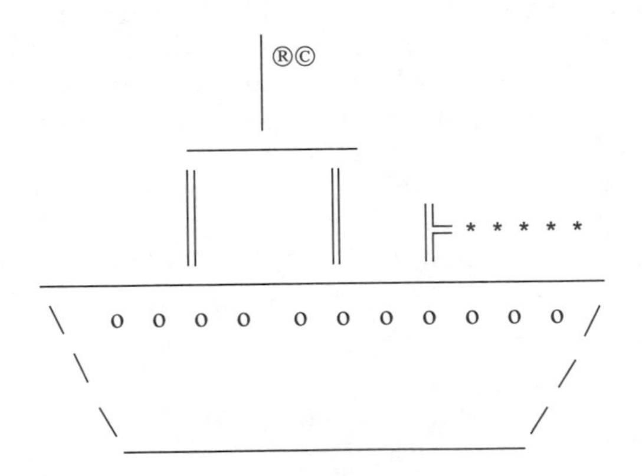

keyboard alone![12] The flag, bearing the letters RC, which I admit is in the wrong place (it ought to be on the stern), will have to stand for Republic of Carpathia.* It was impossible to create a more realistic flag because the only 'flag-like' letters available through the keyboard were R, for 'rights reserved', and C, for 'copyright'. And it may be unusual for a battleship to have portholes, but I couldn't resist putting them in.

Then the coming of the hard disk brought Microsoft's DOS, or 'Disk Operating System', which provided nearly 600,000 bytes. This made it possible to have static colour pictures, with 'a-goblin-is-rushing-towards-you'-type dialogue, giving new force to the old newspaper saying that a picture is worth a thousand words. Six hundred thousand bytes is the length of a long novel.† It represents, in other words, a considerable amount of information, but not nearly enough to support *moving* pictures. An ordinary colour feature film consists of about five billion bytes, the length of 8,000 novels – which may help to explain why it is so much more expensive to make a film than to publish a book. The next step, which gave us games like *The Rose Tattoo* and *The Beast Within*, was the invention of the CD-ROM drive whose disks contain up to 650 *million* bytes, a significant fraction of the information content of a feature film. *The Rose Tattoo*, containing only make-believe figures with actors' voices against painted backdrops, required only one CD-ROM disk. But *The Beast Within*, because it had not only actors but real film, came in six disks.

This required repeated screen messages, which tended to interrupt the player's concentration, asking him to remove one disk and insert another. But then, in the last two years of the twentieth century, came 'Digital Versatile Disks', meaning CD-ROM disks that can hold nearly five billion bytes, *seven times more* than the

---

* I could also have put a lower case w or two in the sky to denote seagulls. But no doubt the seagulls were frightened away by the gunfire. (In early games, changing a lower case w to a capital W gave the appearance of a hostile aircraft rapidly approaching from far off.) The illustrators of early games were of course far more professional than I have been here – within the limits of the tools available to them.

† This number is calculated by counting the total number of characters in the book. But this is only the 'raw' novel. It could not be sold like this in a bookshop. When it has been formatted, its text justified and given suitable fonts, it may contain double this number of bytes.

earlier CD disks.[13] It is obvious where this argument is leading. Bearing in mind the law of Gordon Moore of Intel – 'Moore's Law' – that computer processing power doubles approximately every 18 months, it will be seen that within a few decades it will be possible to run games of trillions of bytes, which for all practical purposes are infinitely complex.

One spectacular improvement, appearing in the latest games, is called 'complete 360-degree mobility'. It puts the player, literally, at the heart of the action. This is a world in which you can turn and move *in any direction*. This is impossible in a theatrical stage set, because if it were attempted there would be no room for the audience. All that a stage set can present is 180-degree mobility. Say that we are in a theatre looking at the stage, and we see people in a drawing room. But it is a stage room. Only one half of it can exist! There can be exits and scenery at the back and to either side, but there cannot be any at the front because this would block the view of the audience. But in a few very modern games, which of necessity take up prodigious amounts of memory, the player/character can move anywhere, giving the impression of absolute immersion in another world.[14]

To prevent boredom and the dangerous effects of idleness, people in starships will pass much of their long periods of leisure time with such unimaginably intricate games. And it will not even be necessary for human beings to go to the expense and labour of creating these games in advance. Computers will create them on demand.

Can computers really create stories in the form of films, in which the player takes a key role? Stories that make sense within their own logical framework, that are filled with puzzles of varying difficulty, and are sufficiently 'character-driven' to be intellectually satisfying? Indeed, to a certain extent they are already doing so.

It is not an insuperable task. Almost every fictional story that has ever been written, from Shakespeare to Agatha Christie, consists of three consecutive parts:

1. The characters are introduced.
2. They strive against one another.
3. The situation is resolved.

Television writers, and a growing number of novelists, people who are stuck while trying to compose stories, often anchor together these three parts with a computer program called Plots Unlimited. It invites the user to choose *one* of the three parts from a huge database, and then weld it to two other parts from the same database.[15] If Shakespeare had been using it, he might first have chosen the line 'ambitious subordinate officer', followed it with 'wins control of the organisation by crooked means', and finally: 'and is removed by fellow-officers who fear his paranoia', and he would have had the plot of *Macbeth*. Plots Unlimited does not *write* the story, but it does help to get rid of the mental block which has left so many writers of fiction staring in frustration at their screens.

The program has obvious limitations. The phrase 'ambitious subordinate officer' does not tell us very much about the actual Macbeth. It gives no hint of his predilection to vacillate and seek advice from witches. For characterisation, there is a program called The Collaborator. This software does not try to construct a story. It asks the writer hundreds of questions about the characters he is trying to create until he knows how they will behave in all conceivable circumstances.[16] This technique provokes another idea. Why do we need characters at all? Why not settle instead for *information about characters*? Baseball fans of the future, it has been suggested, will have no need to attend real baseball matches or watch them on television. Instead, their home computers will generate fictitious games based on statistical records of real players – which will include their individual propensities to get tired, make mistakes, and argue with the referee – and play them endlessly, long after the original players are retired or dead. The fan need not worry about strikes, injuries, and closure of the pitch by bad weather. He will always have a game to watch.[17]

To return to computer games. Can they truly create themselves, so that starship passengers can dispense with vast libraries of disks, not to mention the expense of getting companies back on Earth to create them? It seems very likely. Here is a simple computer program, written in IBM BASIC, which would, if it was developed to the nth degree of complexity, construct a sensible story from an unimaginably huge database of subplots:

```
10 PRINT "You are";

20 RANDOMIZE TIMER
```

```
30 X=INT(RND*10)
40 IF X=0 THEN 30
50 IF X=1 THEN PRINT "imprisoned in a torture chamber."
60 IF X=2 THEN PRINT "fleeing a monster but unable to move."
70 IF X=3 THEN PRINT "in Baker Street, confronting Moriarty."
80 IF X=4 THEN PRINT "in a forest, fleeing a bloodthirsty wolf."
90 IF X=5 THEN PRINT "preaching Christianity to fanatical mullahs."
100 IF X=6 THEN PRINT "in space, falling into a black hole."
110 IF X=7 THEN PRINT "computing infinity in your head."
120 IF X=8 THEN PRINT "in a racing car with two burst tires."
130 IF X=9 THEN PRINT "experimenting with a smallpox virus."
140 IF X=10 THEN PRINT "hoping that what will happen won't."
```

This program is based on a very simple idea. The command in Line 20, RANDOMIZE TIMER, takes advantage of the computer's internal clock, which is continuously 'ticking' while the machine is switched on. At every second of the day or night, it is recording a different number that, provided you are not trying to tell the time from it, for all other purposes may be regarded as random.* The command in Line 30, which identifies the variable X, forces X to be any whole number between 1 and 10. (It also has the option to be zero, but Line 40 tells the program that zero is unacceptable.) Thus, whenever the program is executed, it first displays the message: 'You are', and then jumps at random to one of the ten different 'plots' to make up a complete sentence.

---

* Strictly speaking, the word 'random' should be 'pseudo-random'. A computer, being a man-made machine, cannot produce genuinely random numbers. Only some force of nature, like the impact of cosmic rays from space, can do this. Nevertheless, computer-generated pseudo-random numbers can *appear* to be genuinely random if they are not too closely examined. In the Holmes game, for example, Holmes always seems to make an unpredictable remark to Watson when the pair leave 221b Baker Street to resume their investigation. It's either, 'The game's afoot', 'Back to the chase', or something else chosen by the game's pseudo-random number generator.

This embryo program could be expanded indefinitely. Instead of ten Xs, there could be tens of millions. Each generation of X would take us to a different Part Two ('the characters strive against one another'), along with different sub-plots, and eventually to Part Three, the climax, in which 'the situation is resolved'. The whole story would then be examined – perhaps even electronically – to make sure it contained no step that was unbelievable, absurd, or out of character with the rest. Various puzzles would be inserted, again perhaps electronically, in which the player would be invited to make all manner of wrong choices before deducing the right one, and we would have a game.

When a game is played in a starship, all outward forms of computing machinery can be dispensed with. Why take up valuable space with extra screens, keyboards, mouses, sensitive suits, gloves, joysticks, and all the other gadgets that are used in virtual reality? The game can be played *in one's head.*

We have long and rightly scorned people who claimed to be capable of *telepathy*, of sending and receiving messages without speaking, seeing, listening, smelling, typing, or gesticulating. But we are scornful only because we know that the necessary technology does not exist. Nobody has yet been born with a radio or television transmitter inside his skull, and as for the natural radiation that *is* emitted by the human brain, it does not even contain enough energy to make the lightest piece of paper flutter, let alone send an intelligible message.*

So much, however, for the present. There is nothing in science that bars the future possibility of installing microchips in our brains, devices a micro-millimetre in size that would have all the functions of modern television transmitters and receivers. Long predicted in such science fiction novels as Olaf Stapledon's *Last and First Men* in 1931, Sir Fred Hoyle's *The Black Cloud* in 1957, and William Gibson's *Neuromancer* trilogy of the 1980s, the future technology of 'radiotelepathy' is now being taken very seriously.

The thrust towards it does not come from the computer game industry but from medical research. There are schemes to insert

---

* It *may* be possible that some people claiming to demonstrate telepathy are not performing conjuring tricks and dishonestly concealing the fact. There may, for all I know, be biological mutants who have this ability. But I have never met or heard of such people.

microchips in human brains to gain a better understanding of the architecture of our central nervous system, a technique called *radioneurology*.[18] Such chips will be entirely passive. They will be wholly under the control of the external experimenter. He will use them to explore the brains of his subjects.

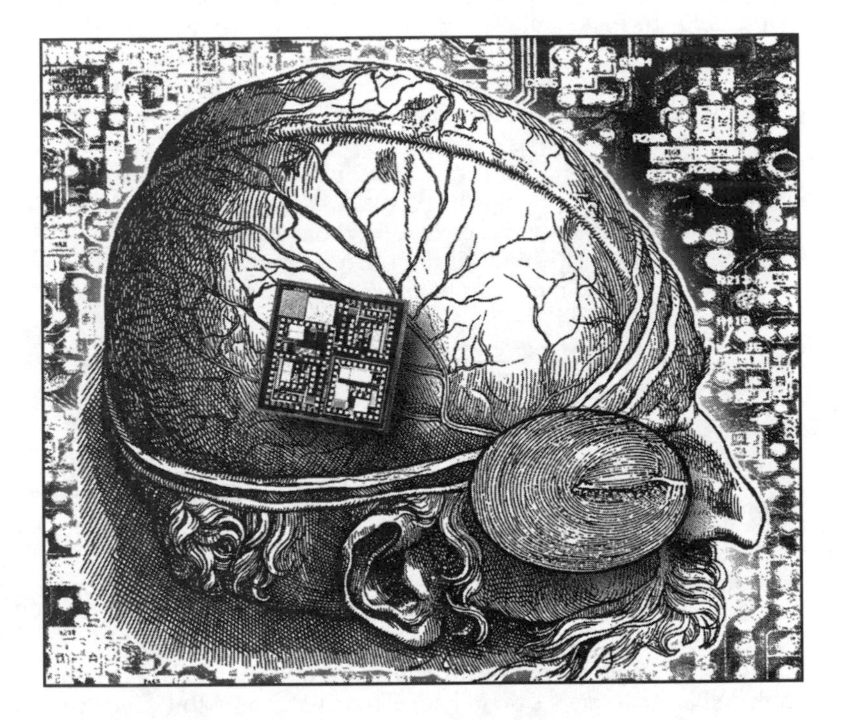

Why should science stop here? There will at last come a generation of neurochips that *can* be manipulated by their wearers, and the era of radiotelepathy will begin, bringing with it, in the words of Freeman Dyson, 'acute questions of ethics'.[19]

It is easy to see how. Many financial houses automatically record all telephone conversations by their executives to discourage insider dealing and fraud. But they can only do this so long as the suspected executives are using telephones, not if they are radiotelepathists! When there is a new generation of people who can communicate with anyone they want, without any sign that they are doing so – and probably using encryption as well – our affairs will be increasingly influenced by a secret elite.* Ethics indeed!

But to look on the more optimistic side, radiotelepathy-driven computer games will surely prevent boredom and dangerous lethargy among the personnel of starships. Instead of being considered a frivolous waste of very expensive time, they must be viewed, because of the fascination they bring, as a productive contribution to the success of the mission. And all the better if one can play them by merely closing one's eyes and activating a skull-borne chip.

Another possibility is that people will play games that have a practical application to their mission – to build a planetary civilisation on arrival. Many such games exist already, although most of them are oriented towards building classical empires on Earth. They are not melodramas, like the games mentioned above. In these more educational games, you cannot be shot, stabbed, bombed, or eaten. You can only succeed or fail at the task of building an empire or nation.

A typical such game is the 1998 *Caesar III*, in which the player's task, as a provincial imperial governor, is to turn an empty landscape into a Roman city. Sites must be cleared of forest so that housing areas can be designated. Roads to these areas must be constructed, otherwise nobody will want to build houses in them. A Senate House must be constructed to accommodate tax collectors. Forts must be built, so that soldiers can protect the city against brigands. There must be adequate food markets, and so forth.

You have to raise sufficient revenues through tax and trade. Otherwise you will only waste money; upon which the Emperor will fly into a rage and order your dismissal. But raise taxes too high, and people will get into a foul mood and emigrate. In all, it is a delicate exercise at compromise between one aim and another.

---

* What makes this prospect even more frightening is that there exists a form of encryption called a 'steganogram', in which not only is a message rendered unintelligible but its *existence* is concealed. A radio or TV message could be made temporarily indistinguishable from natural radiation. When it becomes impossible not only to decipher messages sent between members of the elite, but even to prove that they are sending messages at all, their power over the rest of us will be great indeed!

It seems probable that wise starship commanders will encourage their crews to play games specifically tailored to the tasks ahead of them.*

I do not mean to imply that all the people awake on the ship will spend their time playing games. There will be many who prefer to do science – especially astronomy and cosmology – for which the relativistic effects caused by the ship's high cruising speed will give ample opportunity. Physical exercise will also be important, as will work in preparation for landing. Nonetheless, when contemplating a really long voyage, cut off from the rest of humanity, one cannot neglect the prospects for inflight entertainment.

---

* In 1999, there appeared an exciting and, from the point of view of starships, most appropriate game called *Alpha Centauri*. It is similar in general form to *Caesar III*.

Chapter 17
The Big Sleep

That is not dead that can eternal lie,
and with strange aeons even death may die.

H.P. Lovecraft

I don't want to achieve immortality through my works. I want to achieve immortality through not dying.

Woody Allen

Colonisation of the galaxy, or at least a significant part of it, will be a far more extensive operation than merely establishing settlements on the planets of nearby stars. The Polynesians were not content to settle a few isolated Pacific islands; nor will our descendants be content to do the same in interstellar space.

Rigil Kent, the three-star system that we know alsc as Alpha Centauri – if it proves to have habitable worlds – will be only the beginning. As the technology of starships develops in coming centuries, mankind will want to go ever further afield. There are at least seven stars within an arbitrary radius of 17 light years from the Sun, beyond Alpha Centauri, that *might* possess habitable Earth-sized worlds. Moreover there is one star, CM Draconis at a distance of 57 light years, which was discovered in 1998 to have an Earth-*sized* planet, or a planet only a little larger than Earth. Whether it is habitable of course no one knows. Here is a list of other stars which are possible candidates for having habitable worlds, with their distances.[1]

Star system	*Distance (light years)*
Lalande 21185	8.1
Epsilon Eridani	10.7
61 Cygni B	11.2
Procyon A	11.4
Tau Ceti	11.8
Omicron Eridani A	16.2
70 Ophiuchi A	16.9

Eventually, however, the desire will come to go to much more distant stars, perhaps hundreds of light years away. In such a case, all that I have said in the last chapter about fighting boredom becomes invalid. Because of the extreme difficulties and prohibitive energy costs of flying faster than 92 per cent of the speed of light, such journeys would necessarily take many centuries. Even the on-board slowing of time would be of no use here. Any human beings who attempted to stay awake for such immense periods would be bound to die of old age long before the ship reached its destination.

It will be necessary to put them to sleep. Not with a light anaesthetic, as with a journey like that to Alpha Centauri lasting little more than a couple of years, but by freezing their bodies so that they are for all practical purposes dead, but capable of being revived.

This technique, which has a long way to go before it is perfected, is called 'cryonic suspension', the suspending of life by means of extremely low temperatures.

Crew members would be kept in liquid nitrogen, the temperature of which is not merely frigid. It is *minus* 196 degrees, only 77 degrees above absolute zero, which in turn is the lowest possible temperature in the universe. Atoms at this temperature would no longer move.* (There is no possibility of any lower temperature than absolute zero. You cannot have less heat than no heat at all.) The scale of natural and man-made temperatures goes like this:[2]

* The temperature of a substance depends on the speed – or the potential energy – at which its atoms are moving. I used to imagine as a child that one could heat up a cup of cold coffee simply by vigorously stirring it. But doing this requires the application of considerably more energy than a human wrist can apply!

Where found	*Degrees C*
The Big Bang	Infinite
Interior of hottest stars	1 billion
Core of the Sun	15 million
Hydrogen bomb explosion	50,000
Surface of the Sun	6,000
Iron melts	1,500
Greenhouse effect on Venus	400
Hottest place on Earth	58
Average on Earth's surface	20
Coldest place on Earth	–89
Liquid nitrogen	–196
Interstellar space	–270.45
Lowest temperature of liquids	–272.00001
Lowest temperature of metals	–272.00002
Absolute zero	–273.15

The figures at the lower end of this table represent a truly stupendous cold, temperatures familiar only to laboratory experimenters. Most people have never encountered anything colder than dry ice, commonly found in domestic freezers and ice cream boxes, at a mere minus 78 degrees. This is frozen carbon dioxide, well known from the white mist that surrounds it, caused by the condensation of air.[3]

Nitrogen thus liquefies at a temperature more than two and a half times colder than the coldest it ever gets on Earth, in deep winter at the Pole of Inaccessibility in Antarctica.[4] Liquid nitrogen is neither toxic nor explosive and is the best refrigerant that exists for foods and living organs. Even back in the 1960s, it was routinely used to freeze blood for indefinite storage, and on one occasion it was employed to freeze the still pulsating heart of a dead chicken so that when thawed it resumed a healthy beat.[5]

More recently, several people dying of terminal diseases have, before death, arranged to have their bodies frozen in liquid nitrogen in the hope that, one day, as medical science advances, they can be awakened and cured. And at least six commercial organisations exist to take care of the 'bodies'.[6] The law, in such countries where this technology exists, requires a person to be clinically dead before they can be deep frozen, but after that the process can begin. As one reporter describes it:

> As soon as you are pronounced medically dead, they rush you to a converted warehouse . . . Your circulation is artificially restarted using a heart-lung machine to maintain a flow of oxygen to the brain and to infuse your blood with a biologically compatible substance containing glycerol. This acts as an anti-freeze and averts some of the worst effects of freezing on your brain tissues.
>
> At first cooling is gradual – a big drop in temperature could shatter you – but eventually you are immersed in liquid nitrogen which takes the body to minus 196°.[7]

The patient is medically dead, but may be capable of being revived. It seems that in the future, when this kind of technology is developed – only a gleam in technicians' eyes at present, so don't make a rush for the liquid nitrogen just yet – there will be two different definitions of death: clinical death, when the subject has stopped breathing and his heart no longer beats, and what one expert on this subject, Ralph C. Merkle, calls 'information theoretic death'.[8] The latter term means when it is no longer possible to recover information from the brain *by any means whatsoever*. As Merkle explains:

> It is essential to understand the gross difference between death by current clinical criteria and information theoretic death. This is not a small difference in degree, nor just a small difference in viewpoint, nor a quibbling definitional issue that scholars can debate; but a major and fundamental difference.
>
> The difference is as great as that between turning off a computer and dissolving the computer in acid. A computer that has been turned off, or even dropped out of a car window at high speed, is still recognisable. The parts, though broken or even shattered, are still there. While its short-term memory is unlikely to survive such mistreatment, the information held on disk will survive. Even if the disk is bent or damaged, we could still read the information by examining the magnetisation of the domains on the disk surface. It's not functional, but full recovery is possible.[9]

Reviving a deep-frozen person will be considerably more difficult than extracting data from a smashed-up computer. It will be necessary not only to restore limbs and organs to their former functions, but to revive the brain and all the information it previously contained. To do the latter, one must ensure that all the molecules in the brain are in the right places. Some measure of the complexity of this task may be seen from the estimate that while the average human brain weighs only 1,400 grams, it contains 40 trillion trillion molecules.[10] Any significant damage to this material during freezing, storage, or thawing of the subject will result in true information theoretic death, in which he is either truly dead or a mindless zombie.

But there are reasons for being optimistic. Let us take a look at the current state of *cryonics*, the science – or art – of maintaining people who are legally dead at extremely low temperatures so that they can be revived at some future age.* While the first half of the operation is comparatively easy and has been achieved in many cases, the second half, revival of the frozen person, is far beyond our present technological abilities.

First, the freezing. Wesley du Charme, in his book *Becoming Immortal*, describes how it happens.[11] The body of the 'patient' is first mildly frozen to a few degrees above ordinary zero and placed in a heart and lung machine to minimise any decay of brain cells that might occur between the point of legal death and the beginning of the real freezing process.

Once placed in storage, the process begins, taking several days. Down, down goes the temperature to that of the liquid nitrogen in which the patient is immersed. And there he stays, for years, decades, or centuries. At this temperature, du Charme explains, 'virtually all physical processes have come to a halt. Electrons are barely moving in their orbits. A second's worth of decay at room temperature would take millions of years to occur at minus one hundred and ninety-six degrees.'[12]

The task of reviving the patient at a much later time is impossible today, but can be envisioned in theory.† Any physical

---

* This is not to be confused with the science of *cryobiology*, which merely studies the effects of very low temperatures on living systems, and whose adherents are so strongly prejudiced against cryonics that their professional organisations have been known to expel members for working or publishing in the field.

† It won't be like the scene in the film *Aliens*, in which, after being frozen for many decades, people stagger out of their 'pods', take showers, do press-ups, and are then completely recovered. Nothing even remotely resembling this scene is ever

system, an animal or a machine, can be brought back to its original pristine condition merely by replacing its defective parts. Here we turn again to nanotechnology, the future art of assembling and disassembling molecules under computer control. Nano-devices inside the patient's comatose body will continuously replace all defective cells. The only problem is that as yet we have little idea how to build such tiny devices and make them do what we want them to do without human supervision. But current experimental and theoretical progress strongly suggests that these feats will be possible within a few decades.[13]

Even without nanotechnology, however, living tissue can survive for millions of years, and without even being frozen. Setting aside the underlying plot of *Jurassic Park*, in which fictitious scientists retrieve dinosaur DNA from dead insects captured by tree resin, the DNA of magnolia leaves which grew at least 17 million years ago has been found largely undamaged at the bottom of a lake.[14] If nature can do this, so, surely, can human technology. And we hope only to preserve life for centuries or millennia, not for tens of millions of years.

Manned starships with everyone on board deep-frozen will thus be the only way – barring the development of faster-than-light travel – to colonise the planets of the remoter stars in our region of the galaxy. Robots will put the human travellers to sleep, and robots will awaken them months, or perhaps years, before they reach their destinations. Robots will navigate, and robots will manage the ship and all its operations. No human will stay awake to supervise the work of the robots, and the success of these age-long journeys will depend only on the quality of the ship's construction, the intelligence of the robots, and their imaginative resources in an emergency.

The bad news, of course, is that anything that can go wrong will go wrong. There will be many voyages in which, due to some faulty computer programming, the revival technology fails, and the ship cruises endlessly through the darkness of interstellar space, with the bodies and brains of its crew rotting beyond

---

likely to happen. There would be fractures in their tissues caused by thermal strain. If the newly thawed-out subject tried to walk, says Merkle, he'd 'fall into pieces as though sliced by many incredibly sharp knives'.

recovery. Or the supervisory robots may malfunction. They could make an error in navigation by selecting the wrong guide star. In either case, the ship would travel for ever, probably never coming near a planet because the distances between planets will be incomparably vast. Yet judging by Earth's history, the prospect of such occasional losses will be acceptable. They will no more affect future migration than did the disappearance of ships in the Atlantic.

But the good news, when things do *not* go wrong, will be economic. There will be enormous savings in costs compared with the expense of shorter journeys. These long-range ships will be far smaller than those going to nearby stars like Alpha Centauri. There will be no need for luxuries or even comforts; robots do not require them. Nor will there be any need for food, water, entertainment, or circulating air. The passenger and cargo compartments need only be large enough to accommodate the stacked bodies of the frozen people, the machinery needed to keep them frozen and to monitor them, the robots themselves, and the heavy equipment the colonists will need on arrival.*

I spoke of the remoter stars 'in our region' of the galaxy. For it must be remembered how vast our galaxy is. On the scale of the whole universe, it is an infinitesimal speck. But on the human scale, it is gigantic almost beyond our ability to imagine. It contains at least 250 *billion* stars. There are nearly 10 trillion kilometres to a light year, and 28,000 light years lie between us and the galaxy's centre. Beyond that, to the opposite edge of the galaxy, is another 50,000 light years.

Unless they are especially bright, or lie in beautiful clusters or nebulae, we do not even trouble to name the stars that lie more than a few hundred light years away. Only 563 stars in the entire galaxy have actual names; the rest have merely numbers or letters.[15] If they are among the 24 brightest stars of a particular constellation, we name them, in order of brightness, after the Greek alphabet, from alpha to omega. But when they are less bright than that, they just get a number preceded by a letter,

---

* This being so, it might be asked: why not deep freeze the crew on shorter flights as well? The answer is that it will be too risky and not necessary. Freezing people is such a complex technology that there is no point to it except for very long periods.

chosen at different times from several different cataloguing systems.* All this, however, will change over the coming millennia, as more distant parts of the galaxy are colonised. If deep-freezing and revival technology prove successful, a time will come when there are as many stars with actual names as there are now named towns on Earth.

---

* It is always a great mistake to give money to organisations that promise in return to name a star after you. They cannot do this. The star name will be recognised by no one, and the people who took your money will laugh all the way to the bank. The International Astronomical Union alone can give names to celestial objects that will be universally recognised.

Chapter 18

The Fading of the Stars

> Frodo looked and saw, still at some distance, a hill of many mighty trees, or a city of green towers: which it was he could not tell. Out of it, it seemed to him that the power and light came that held all the land in sway. He longed suddenly to fly like a bird to rest in the green city.
>
> J.R.R. Tolkien, *The Fellowship of the Ring*

> If comets pervade our entire galaxy, then the galaxy is a much friendlier place for interstellar travellers than most people imagine.
>
> Freeman Dyson, *Disturbing the Universe*

In space, as at home, there are times when comfort is the enemy of progress.

Recall the Earth-sized world that I described in Chapter 1. It seemed to have everything that colonists would need. It was in a Goldilocks orbit (neither too hot nor too cold), circling a Sun-like star, with abundant water, oxygen, and plant life. After a few lifetimes of diminishing hardship, its inhabitants would enjoy standards of living far higher than anything they had possessed in the solar system.

It would be a pleasant existence, perhaps too pleasant, for there would be a danger of decadence. This seems a particular peril, more subtle than the threats of environmental ruin or civil war, but just as real, since mankind, as a species, is prone to the temptations of lethargy. We can imagine how, as generation succeeded generation, the colonists became so proficient at taming their wilderness that everything they desired became instantly available. Their world would come to resemble the Island of the Lotus-Eaters where Odysseus and his shipmates became so contented with their surroundings that they almost lost the will to respond to new challenges.*

However, such challenges, necessary for the psychological and material survival of the colonists, will be present in the form of other habitable celestial bodies that seem certain to exist in almost boundless profusion in most stellar systems.

I refer to comets.

I must briefly explain this. Until the middle of the eighteenth century, comets represented the greatest mystery of the night skies. What were these fiery invaders, where did they come from, and what did their appearances *mean*?

To Aristotle, and hence to all thinkers until the scientific enlightenment (who believed that the only way to solve a problem was to look up what Aristotle had written about it), these questions had simple answers. Comets could not come from the heavens because the heavens were perfect. In a perfectly ordained cosmos there was no place for unpredictable apparitions. They were therefore supernatural emanations of the terrestrial air that came to foretell disasters, usually the death or overthrow of important people, as in Shakespeare's *Julius Caesar*:

> When beggars die, there are no comets seen;
> The heavens themselves blaze forth the death of princes.[1]

Then came the astonishing solution to the mystery of the object that was to become known as Halley's Comet. All the other comets seemed to come and go unpredictably, but not Halley's. It put in roughly regular appearances. Approaching the Sun approximately every 76 years, it has been seen 30 times between its earliest recorded appearance in 1058 BC and its most recent one in 1986.†

---

* As Tennyson described this island through the eyes of its victims in his poem 'The Lotus-Eaters', 'A land in which it seemed always afternoon.'

† There is an apparent discrepancy here. Its period of 76 years ought to have made it appear 40 times between these two dates, not 30. But a comet's period is only approximate. Being fragile 'dirty snowballs', comets are highly sensitive to the

Many other comets with shorter periods have appeared and reappeared, but the astronomer Edmund Halley, alone in the seventeenth century, suspected that people were repeatedly seeing the same objects.[2] He surmised the true nature of comets: that they were regular visitors from far out in the solar system. After the appearance in 1682 of the comet that was later named after him, he calculated its period from what he believed were its previous ones, predicting its reappearance in 1758.

It duly appeared, a year late, in 1759, an event commemorated by a splendid painting by Samuel Scott – showing people gazing at it in amazement from the Thames Embankment in London – and confirming the long-dead Halley's belief that those comets that swept past Earth all came from the same mysterious region in the distant reaches of the cosmos.[3] Halley's Comet became – and has remained – the most famous of all comets. As an anonymous contemporary doggerel put it:

Of all the comets in the sky
There's none like Comet Halley.
We see it with the naked eye,
And periodically.[4]

But where and what was this 'mysterious region'? Somewhere out there, reasoned Jan Oort, one of the greatest astronomers of the twentieth century (the discovery of the shape of our galaxy and its dimensions is largely his work), must lie a vast cloud of comets.[5]

To be still undetected by telescopes, this cloud must be very far away. It must also be unimaginably huge. Consider. One or two new comets invade the solar system each year, although most never become bright enough to attract general attention. When Oort reflected that these invasions had been going on for the billions of years during which the solar system has existed, he realised that the total number of comets surrounding the solar system must be colossal.[6] Indeed, it is now generally believed that the great Oort Cloud of comets is a spherical cloud surrounding the solar system with a radius of up to a light year, one quarter of

---

gravitational pull of planets, and cannot be expected always to be punctual. And on many occasions, no doubt, when people were preoccupied with killing each other, its visits may not have been recorded.

the distance to the nearest star. And it is thought to contain no fewer than *10 trillion* comets (10 followed by 13 noughts), with a total mass three times that of Earth.[7]

There is no reason to believe that the Sun is unique in being surrounded by a huge sphere of comets. For the Sun is considered in almost every respect an average star. All other Sun-like stars, therefore, are likely to have their cometary clouds. If this is so, comets pervade the entire galaxy and can rival planets as places where humans can establish settlements.

It is possible to live in an Oort Cloud of comets, either the one surrounding this solar system or another. The total area of the Oort Cloud is believed to be about 1,000 times that of Earth's.[8] Each comet will be an island in space, separated from its neighbours by no more than a few million kilometres.

These 'islands' will be significantly larger than the comets we have observed during their invasions of the inner solar system. For they will never have had close encounters with the Sun in which to lose any of their icy material. Halley's Comet, for example, loses more than 200 million tons from its nucleus every time it comes close to the Sun, and in 150,000 years' time, after about 2,000 more sunward visits, it will have entirely melted away.[9]

Europe's *Giotto* spacecraft, which intercepted Halley's Comet in 1986, found a peanut-shaped object measuring 15 × 8 × 8 kilometres. But bear in mind that it is likely to have been broiled by the Sun countless times before Chinese astronomers first saw it 3,000 years ago, and that when some unknown event first disrupted its original orbit in the Oort Cloud and drove it sunwards, it must have been many times its present size.

The average comet in the cloud may therefore be about 50 kilometres wide, giving it a surface area of 8,000 square kilometres, about the size of a small English county.[10]

Such mini-worlds could accommodate tens of thousands of people. Freeman Dyson has proposed that trees could be grown on the surfaces of comets.

Not trees as we know them, but artificial trees that would be part plant and part machine, a mixed creation of DNA and computer software. Instead of mainly synthesising cellulose like terrestrial trees, their cells could make pure alcohol or octane or any other desired organic chemical.[11] Rising from the weak gravitational fields of cometary surfaces, they could be giants thousands of metres high, making Earth's tallest trees, the Californian redwoods, seem like diminutive shrubs in comparison.

Factories and homes would exist in their myriad branches, a situation reminiscent of the arboreal city of Lothlorien, in Tolkien's *The Lord of the Rings*, where palaces are built high up in the trunks of huge trees. Electric power would be barely available from the faint and far-off Sun (or parent star, depending where in the galaxy they might be). But there would be plenty of local materials for creating energy. Even a comet as small as 10 kilometres wide will contain more than six million tons of deuterium, the naturally occurring heavy isotope of hydrogen.[12] Such supplies on a single comet would be enough to sustain a community of thousands of people for millions of years with the energy of nuclear fusion.

And once life had established itself on a comet, its bacterial spores would drift through space, driven by stellar winds, until they found another. The time would come when colonists seeking suitable comets might find that the necessary biological infrastructure was already present and waiting for them. Dyson sees the self-seeding of comets with plants as powerful a mechanism for spreading life across space as fast starships. It will soon increase beyond the possibility of control:

> Whether or not comets provide convenient way stations for the migration of life all over the galaxy, the interstellar distances cannot be a permanent barrier to life's expansion. Once life has learned to encapsulate itself against the cold and the vacuum of space, it can survive interstellar voyages and can seed itself wherever starlight and water and essential nutrients are to be found.
>
> Wherever life goes, our descendants will go with it, helping and guiding and adapting. There will be problems for life to solve in adapting itself to planets of various sizes or to interstellar dust clouds. Our descendants will perhaps learn to grow gardens in stellar winds and in supernova remnants. The one thing that they will not be able to do is to stop the expansion of life once it is well started. The power to control the expansion will be for a short time in our hands, but ultimately life will find its own ways to expand with or without our help. The greening of the galaxy will become an irreversible process.[13]

This, then, will be the challenge that the colonists of planets will have to face to prevent their new world from becoming a land of Lotus-Eaters. The comets that surround their parent sun will be their challenge and their workshop. Above all, they could be factories for the construction of new starships, a task that because of gravity, and the danger of ruining the environment, could be almost impossible on the surface of a planet.

Comets, also, like the traditional frontiers of adventure, will always provide an outlet for the colony's most ambitious and energetic members. Wherever our descendants find planets, these lesser celestial bodies will exist in countless abundance. They will be the cosmic equivalents of the Wild West or India's North-West Frontier. Their psychological influence will be decisive. A planetary colony may perish in its early generations from failure to develop and organise itself. But it will never expire in later, more prosperous days through inertia and sloth. The natural conditions of the universe dictate otherwise.

Cometary life will carry with it another consequence. In a more remote future, it will be possible to tell from distances of hundreds of light years what parts of the galaxy are occupied. It is uncertain whether Dyson, in predicting the 'greening' of the galaxy, has named the correct colour. It may be that the light of the stars, obstructed by the artificial 'shells' erected around a myriad comets, will have dimmed, turning towards the infrared.[14] The comets themselves will remain invisible, but the stars that they surround, when viewed from afar, will seem to have faded.

I pointed out earlier that stars, as seen by the crew of a fast relativistic ship, will turn blue. But this is an illusion caused by high speeds. Here, the prediction is that the stars will literally change colour.

Imagine that we look today at some distant star cluster through every kind of astronomical instrument that is available. We see and hear nothing that would suggest biological activity. Except when they are obscured by natural dust clouds, the stars are white, harsh, and unblinking, and through the radio spectrum comes only meaningless static.

But look again at this star cluster after a suitable interval, say one of several thousand years. No longer are its stars white and unblinking. They are detectably reddened and dimmed. And no longer are they dead and silent. Every one of their inhabited planets will emit a vast radio 'noise', vastly exceeding the natural radio output of its parent sun. This already happens with Earth, whose artificial radio activity is detectable out to a distance of about 50 light years, a radius that has been increasing by a light year every year since the first use of ultra-high-frequency television allowed signals to escape into space.

While changing the colour of the galaxy – or at least part of it – man will make it continually more noisy in the radio spectrum. Such will be one of the most noteworthy achievements of Interstellar Man.

Another will be the eventual interconnection by radio, first of the whole galaxy, and then of a big part of the larger universe.

Chapter 19

A Cosmic Internet

> Every time I receive a cablegram in code, I have the same feeling of pleasurable excitement. There is the familiar envelope lying on my desk, marked 'Cablegram: Urgent'. I rip it open and discover the single mysterious word BIINC. The message is from our Venezuela office. Visions at once loom of secret documents, beautiful women, and dark Latin American intrigue. Then I turn to my code book and find BIINC: *What appliances have you for lifting heavy machinery?* This sort of thing can be very debilitating.
>
> Jack Littlefield, 'Melancholy Notes on a Cablegram Code Book', *The New Yorker*, 28 July 1934

The picture I have given of Interstellar Man is in some ways depressing. Thousands of interstellar communities are envisaged, each presumably self-sufficient, but because of the limiting speed of light it may seem that only those very close to one another would be able to communicate within less than a human lifetime.*

This fact, long obvious to scientists, became widely known to the general public in 1974 when the astronomer and alien life

---

* This statement is only true, of course, if our descendants do *not* discover a means of travelling instantaneously through space. (See Chapter 13, 'Faster Than Light?'.) Such people would be living in a universe with different physical rules, and none of the arguments in this chapter would apply to them.

enthusiast Frank Drake sent a famous radio message off into space. It was a remarkable exploit. Hoping to establish contact with an alien civilisation, he used the giant radio telescope at Arecibo, Puerto Rico, to direct a signal to the globular cluster of stars known as M13 in the constellation of Hercules. (The name means that this is the thirteenth of a catalogue of 103 unusual objects published in 1781 by the French astronomer Charles Messier.)

Both the choice of the destination cluster and the nature of his signal were ingenious. Globular clusters are excellent places to look for alien civilisations. The stars in these clusters are the oldest in the universe, having existed even before galaxies. Being so old, and also very numerous – M13 contains about half a million stars – the chances of the signal one day being read by alien beings (if any exist) are quite good.

The signal consisted of 1,679 on–off pulses. Why this number? Because, as any numerate alien would know, it is the product of the two prime numbers 23 and 73. Drake then drew a 23 by 73 pictogram with pictures in binary code showing what humans look like, the structure of our DNA, the five chemical elements most important to life, the appearance of the human body, the location of Earth, and so on.[1]

But the unfortunate side of this experiment, dampening the enthusiasm of the alien-hunters, is that it will be extremely slow to yield results. The M13 globular cluster is approximately 24,000 light years away. And because all radio messages in space move at the speed of light, no slower, no faster, at least 48,000 years must elapse before we can get an answer from M13 to Drake's signal.* Trying to have conversations with aliens – who may or may not exist – is today just as frustratingly slow a process as communications between human civilisations will be within a few millennia, and for exactly the same reasons.

And not only would it take 48,000 years to send a message and get a reply from beings who lived in M13. For such a message to

---

* Not everyone applauded Drake's project. The late Professor Sir Martin Ryle accused him of endangering the human race by advertising our existence to possibly aggressive and hostile aliens. It is possible, however, that Ryle's fears may come true within a much shorter time than 48,000 years. For there is a Sun-like star 68 light years from us, Eta Herculis, almost in line of sight of M13, which might carry a hostile civilisation that could pick up Drake's signal. A crisis could therefore be upon us by 2110!

cross the galaxy and back again would take 200,000 years, and the round-trip message time to our large neighbouring galaxy in Andromeda would be a wearisome four million years. But is there any way to speed up this process without breaking any known laws of physics, without, that is to say, exploiting the 'faster than light' universe? Surprisingly, there might be.

Timothy Ferris, in his 1988 book *Coming of Age in the Milky Way*, proposed what could be called a galactic Internet. Without involving faster-than-light signalling, it would vastly decrease the time it took to communicate across tens of thousands of light years, between ourselves and aliens – or, in the far future, between isolated human communities. Each community would, in effect, have its own website.

First, however, let us see how this remarkable idea would work on the much smaller scale of our own solar system. This would make it easier to see how it could be expanded into much vaster regions of space. The space agency NASA is currently preparing to 'wire up' the solar system by creating a network of spacecraft that will provide an infrastructure for communication and navigation by future explorers.[2] The first planet to be thus 'hooked up' will be Mars, an obvious choice in view of plans to send people there in the first part of the twenty-first century. Starting in 2003, the agency wants to surround the Red Planet with about six navigation-cum-communication satellites. They will both tell human explorers exactly where they are on the Martian surface and handle all communications between the two planets.

Why, it may be asked, should *satellites* be needed at all for such communications? Why should astronauts on Mars not simply talk to Earth through a transmitter on the surface of Mars? The answer is that they cannot always do this since all too often Mars itself will be in the way. Earth, in other words, would be over the horizon. But this problem does not arise if the satellites are well clear of the planet. They would be at altitudes of 17,000 kilometres above the equator – the equivalent orbit of Earth's geostationary satellites which always stay above the same point on the surface.* They would stay in contact with Earth all the

---

* These orbits have been named *areostationary*, after Ares, the Greek god of Mars. Because of Mars's smaller size, they are slightly less than half the 36,000 kilometres of Earth's geostationary orbit.

time except when Earth and Mars were on opposite sides of the Sun.

Eventually, this system could be expanded to other of our local planets and moons. Wherever human outposts existed, whether manned or robotic, from Mercury to Pluto, there would be high-altitude communication satellites. As one writer foresees, 'Future explorers wanting to phone home will simply tap into this "interplanetary Internet" to communicate anywhere in the solar system.'[3]

Now imagine a similar scheme, as Ferris sees it, on a galactic scale. It would work like this. Instead of putting satellites round a planet, each human community would send out an automated transmitting/receiving station *into interstellar space*. It would be placed in a 'parking orbit' in some neighbouring star system, where it would act as a hub of communication. This station, as I suggested earlier, would be like the website of its creators. It would carry all information about them that they wished the rest of the galaxy to know. Serving both as a computer and a library, the station would even be programmed to make replicas of itself; and in turn place these in other parking orbits in other neighbouring star systems.*

Several tens of millennia hence, it will be far from clear what star systems are inhabited and who is living where. Records will exist, but may be inaccurate through failure to keep them updated. Perhaps humans, some groups of them at least, will wander through the centuries like nomadic tribes in some vast wilderness. In some cases, they may not even want to have their movements and migrations tracked. But they will want to keep in touch. Wherever they wander, they will always need information about the rest of humanity. It therefore seems likely that all stations will continuously – and automatically – broadcast information in the direction of *likely* sites of human life while, at the same time, listening for and recording all incoming data.

---

* There is nothing fanciful about the idea of computers one day being able to make identical copies of themselves. The mathematics of such a procedure was laid down in 1948 by the computer pioneer John von Neumann in a lecture called 'The General and Logical Theory of Automata'. Five years later, in 1953, when Francis Crick and James Watson uncovered the structure of DNA, the core of all life, they found that it was essentially the same as the sequence for machine reproduction proposed by von Neumann.

Eventually, says Ferris, 'everyone is receiving and sending data to and from all the other worlds through these local junction terminals, which may be in their own star system or the one next door.'[4] Conversations could be speeded up still further. Each station could be made to replicate itself several times until there was a huge hub of them spanning the galaxy.

With this network in position, Ferris estimates, we might wait no more than a century to get answers to messages that typically in an *un*-networked galaxy might take 100,000 years. And all this because people started talking without knowing for certain to whom they were speaking. For those seeking it, the information was already there, waiting to be downloaded.

Superficially, one could compare this system with the messaging system in a company's internal computer network. Every employee has a computer terminal on his or her desk, and can use it to send an electronic message to any other. It makes no difference that intended recipients of messages are away from their desks or on holiday. The message will be in the computer, waiting to be downloaded. But there is a huge difference between this and what Ferris is proposing, not just in scale but in principle as well. At regular periods, the company must erase all its old messages because its computer has insufficient memory storage space. (True, office computers of the future will have much bigger storage capacities, but their memories will still be finite and will still need periodic erasures.) In the boundless vastness of space, however, there will be unlimited facilities for storing data. Entire asteroids, bodies 30 or 40 kilometres long, could be adapted for this purpose. No information that the network recorded need ever be thrown away.

It is interesting also that tiny amounts of data can generate vast amounts of information. Before the coming of computers, companies used to communicate by cablegram, and the longer a cablegram was, the more expensive it was to send. And so, to reduce this expense, a new communication system was invented, inherited from the older system of naval signal flags. Single code-words, usually of five letters, were sent that conveyed a much longer message. As in the story of Jack Littlefield with which I began this chapter, a manager had only to reach for his code book to discover the true meaning of a five-letter message. These codes were not intended to guard secrets, merely to save money. (Designed merely to streamline communication, they are not to be confused with the ciphers I mentioned in Chapter 3, 'The Twilight of the State'.) Here are some samples of codes and their meanings from a typical commercial code book of the 1930s:[5]

AHXNO. Met with a fatal accident.
ARPUK. The person is an adventurer. Have nothing to do with him.
AROJD. Please advertise the birth of twins.
BUKSI. Avoid arrest if possible.
CULKE. Bad as possibly can be.
DEOBI. A great battle is now raging here.
ELJAZ. Will have to get the bottom examined before proceeding.
EWIXI. Very few cases of cholera are now reported.
LYADI. Arrived here with decks swept, boats and funnel carried away, cargo shifted, having encountered a hurricane.
NARVO. Do not part with the documents.
OAVUG. An epidemic of foot-and-mouth disease has broken out here.
OBNYX. Escape at once.
PUMZI. Can you combine horses and grain?
PYTUO. Collided with an iceberg.
SKAAE. Unreadable, the writing having been obliterated by water.
URPXO. For what use was the mixing machine intended?
WUMND. Have every reason to believe oil will be struck.
ZEIBI. Arrest all passengers.

Most commercial code systems in those days would contain about 50,000 codes. Code books (they must have been bulky volumes) were divided into two halves, one for the translation of code words and the other for finding the right code words with which to compile a message. Armed with this book, one could read any incoming message or create any outgoing one.*

A system like this could be used on a galactic scale. It would

* Provided, of course, that there was a code to match the message one wanted to send! Naturally, code books of the 1920s and 1930s had no codes that related to transatlantic air travel. The books had to be constantly updated as new technologies emerged.

enormously speed up communication by avoiding repetition and verbosity, and putting lengthy pieces of text and graphics into shorthand. If you want to know what the people of Planet X, 100 light years away, are thinking, you do not have to wait two centuries to find out. Your station merely has to pick up the information about Planet X that has already been stored in its station and neighbouring ones, a procedure that may take no more than two or three years. As in today's Internet, no one will have to make more than what amounts to a local phone call.

Thus, run by intelligent computers that constantly increased their knowledge from the information they acquired, the network would be immortal. It would continue to exist and grow long after the civilisations that created it had become extinct. In the long run, Ferris adds, 'it might evolve into the single most knowledgeable entity in the galaxy. Growing in sophistication and complexity with the passage of eons, for ever articulating itself among the stars, it would come to resemble nothing so much as the central nervous system of the Milky Way.'[6]

As mankind expanded ever more deeply into the galaxy, he would create new transmitting and receiving stations. Each such station would be like a website for that community, its data constantly being enlarged. Far in the future, perhaps millions of years hence, the network may have spread beyond the Milky Way into other galaxies. As Ferris puts it:

> Intergalactic question and answer times go to many millions of years – too long for mortal beings to wait, but perfectly manageable for an interstellar network. The network could afford to fashion giant antennae, use them to broadcast powerful signals to the Andromeda galaxy, even to the populous heart of the Virgo supercluster of galaxies, 60 million light years away, and then wait for a reply.
>
> Every world on every network would stand to benefit as galaxy after galaxy established contact, spinning electromagnetic threads across the expanding universe and exchanging the wealth of galactic libraries. The human species is only about two million years old, a time equal to that required for a message to travel one-way from the Andromeda galaxy to ours. But if information about the Andromeda galaxy and the history of its worlds were already stored in our galaxy's network, we might be able to begin accessing it within a matter of decades after making contact.[7]

This suggests to Ferris the evolution of a 'cosmic mind', a universe brought to life by human communication and thought.[8] But in the long term – in the *very* long term – the mind itself will face extinction.

Chapter 20

Extinction and Eternity

'Such,' he said, 'O king, seems to me the present life of man on Earth, as if on a winter's night when you sit feasting with your ealdormen and thegns, a sparrow should fly swiftly into the hall, coming in at one door and instantly flying out through another. While indoors it is untouched by the winter's fury, but this calmness being passed almost in a flash, from winter going into winter again, it is lost to your eyes. Somewhat like this appears the life of man.'

The Venerable Bede,
Ecclesiastical History

Think of a ball of steel as large as the world, and a fly alighting on it once every million years. When the ball of steel is rubbed away by the friction, eternity will not even have begun.

David Lodge,
The Picturegoers

A dungeon horrible on all sides round
As one great furnace flamed, yet from these flames
No light but rather darkness visible.

John Milton,
Paradise Lost

It is now necessary to look much further into the future, to an age when interstellar travel will no longer be a voluntary option. For the time will eventually come when man will be *compelled* to use starships if he wishes to survive. The Sun, which makes life on Earth possible, will not always behave in the benign way that it has done throughout the history of civilisation and far back beyond that. It will turn hostile, and our descendants will have to escape from it.

But this will only be the beginning of the travails of interstellar man. For during a timescale far greater than 99 per cent of the lifetime of the universe, there will be no stars to travel to. They will all have burned out because there will be no more atoms with which to replenish them, and the cosmos will remain for ever dark, empty, and meaningless.

With this further revelation, the argument for interstellar travel changes yet again. It becomes aesthetic rather than commercial or practical. In the vast timetable that I am going to describe, the 'window' for exploring the cosmos will appear as incomparably tiny. The universe that we know will only be there to be explored for a fleeting instant. It would be a cosmic tragedy if man and the universe both became extinct without the one having ever acquired first-hand knowledge of the other. To see why this must be so, there is a need to examine this grim view of the long-term future.

First the nearer future. Long before the Sun turns hostile, people on Earth will face formidable perils from another source. Danger will come from the galaxy; not from alien invasions (although that cannot be discounted), but from the galaxy's physical nature.

For the past 100,000 years or so, the solar system has been passing through an environment that appears to have been exceptionally benign. The space through which it has travelled has been empty of the clouds of the gas with which the galaxy is filled, and essentially *clean*. This, however, cannot last. Within a few tens of thousands of years, the merest fraction of a second on a cosmic timescale, we will be entering a cloud of interstellar gas that has been called the 'Local Fluff'. Some astronomers go further and call it a 'doomsday cloud'. It will interfere with the solar wind that normally protects us from cosmic rays, high-speed particles that come from violent events in the depths of the universe. With this protection removed by the invading gas, cosmic rays will rip through our atmosphere, destroying the molecules that form life. No one knows how serious the effects will be, but there is a strong chance that Earth and its neighbouring planets will cease to be healthy places to live.[1]

But much worse is to come later – much later. Our Milky Way galaxy is a *spiral*, which means that if looked upon from above it would resemble a Catherine wheel. Interspersed with its comparatively empty regions, there are immense spiral arms consisting of all manner of debris: gas and dust that one day will become stars, black holes, bloated stars that are due to explode as supernovae, and stray lumps of iron and nickel from such explosions in the past.

The entire galaxy rotates, taking about 200 million years to make a complete circuit. But unfortunately for our remote descendants, the spiral arms rotate *in the opposite direction to that of the galaxy itself*. Thus, every so often, since the Sun and its planets rotate with the galaxy, they rush into one of the spiral arms with the speed of a head-on collision.

Earth in the past has suffered many 'mass extinctions', in which much of its animal life has simply been wiped out. It turns out that at the time of all these extinctions, the Sun was inside one of the spiral arms. It can only be surmised that, being inside a much denser region of space, the Earth was frequently and lethally bombarded by debris.

Let us look at the record. Sixty-five million years ago, when most of the dinosaurs perished, apparently from the impact of a giant asteroid, we were in the heart of the Sagittarius arm (so named because of its present direction in the constellation of Sagittarius).

Earlier, both at 212 million years ago and 245 million years ago, there were two more mass extinctions. Nobody knows their immediate cause, but it may be no coincidence that during both these catastrophes we were in the middle of the Scutum arm.* Another such mass extinction took place half a billion years ago, when we were in the Norma arm. We are now in an empty region of the galaxy. But we are headed towards the Perseus arm, a passage that seems certain to bring further mass extinctions, due to begin within another 140 million years.[2]

---

* From the human point of view, it is perhaps mistaken to call all these events 'catastrophes'. If the dinosaurs had not been wiped out 65 million years ago, we would probably never have come into existence. All one asks of a catastrophe is that one's own species should not be on the receiving end of it.

The next peril will be an increasingly hostile Sun. Let us see how it will be possible to escape from it. Here, temporarily, the reason for building starships will be a matter of life or death, not aesthetic. Over a very long timescale – but still a timescale that we can imagine without too much mental difficulty – our parent star is growing hotter and brighter. This has nothing to do with the 11-year cycles in which its radiation fluctuates back and forth by a fraction of 1 per cent, very slightly affecting the Earth's climate, but a far greater change that occurs over those epochs lasting hundreds of millions of years which are known as 'geological time'.

The Sun is now a third brighter and hotter than it was soon after its birth as an ordinary star some five billion years ago.[3] It cannot for ever remain constant because more and more of its nuclear fuel – hydrogen – is being used up.

Three-quarters of the Sun is made of hydrogen. At present there is a vast difference in temperature between this gas in its outer layers and in its core. At the surface it is a 'mere' 6,000 degrees, while in the core, the Sun's nuclear furnace, where four million tons of hydrogen are being converted into helium every second, temperatures rise to 15 million degrees.[4] But the Sun behaves as if it had a built-in determination to carry on its nuclear reaction regardless of cost. And cost there is! As more and more hydrogen in the core is converted into useless helium ash, more and more hydrogen from the outer layers pours into the core to keep the reaction going.

And so, while an ever-increasing proportion of the Sun's mass is turned into active fuel, the temperature of the whole star must inexorably rise.

It used to be thought, in the last century, before the principles of nuclear fusion were understood, that the opposite would happen, that the Sun would get colder as it burned out, just as a log fire gets colder when its logs are turned into ash. In H.G. Wells's great novel of 1895, *The Time Machine*, the Time Traveller moves countless millions of years into the future in his marvellous contraption, and surveys the scene around what once had been London:

> I cannot convey the sense of abominable desolation that hung over the world . . . The sky was no longer blue. North-eastwards it was inky black, and out of the darkness shone

> brightly the pale white stars. Overhead it was a deep Indian red and starless, and south-eastward it grew brighter to a growing scarlet where lay the huge hull of the sun, red and motionless.[5]

But cold Sun or hot Sun, it will make no difference. When the Sun dies, man's descendants will have to leave its vicinity, and eventually the solar system. The Sun will turn hostile in just over 1,000 million years from now. It will grow so hot that the oceans will evaporate. There will be no more rain, and all plant life will die. In this desert, there will be no possibility of agriculture, and anyone who elects to stay on the planet will starve.[6] Starships will no longer be the instruments of financial schemes but a necessity of life. Assuming that large numbers of people still inhabit the Earth, this may prove a difficult proposition! It may be hard to imagine starships that could contain billions of people, or alternatively billions of starships each of which could carry scores of people. But the choice will be simple: migrate or perish.

There may be an alternative strategy, proposed in 1996 by Lorne Whitehead of the University of British Columbia. It is to *move the Earth away from the Sun* to a more distant orbit where conditions would remain more favourable for a reasonable amount of time. It would be very difficult to move the Earth directly, but the same effect could be accomplished, Whitehead suggests, by moving the Moon with antimatter rockets, since the Earth will always follow the gravitational pull of the Moon.[7]

The Moon's nearside surface could be covered with a trillion high-pressure argon arc lamps that simulate sunlight. (Such lamps are already being manufactured for other purposes, by Vortex Industries in Vancouver.) There would be one lamp for every nine square metres of the nearside lunar surface. With this number, Whitehead calculates, each person on Earth – assuming a population similar to today's – would receive the radiation of 200 lamps. Erecting and powering them might seem an expensive project, but not if everyone's lives depended on it.

Thus, since the Moon is the same apparent size as today's Sun, it would simulate the appearance and warmth of the Sun. If this enterprise was executed promptly, *before* the heating up of the Sun evaporated the Earth's water, mankind would have a perfect imitation Sun. The only difference would be that the tides would be a third weaker, because of the real Sun's absence.

'Everything on the planet would look the same,' says Whitehead. 'There would be golden sunshine, blue sky, fleecy clouds and rainbows.'[8] It is interesting that other scientists feel the same optimism about mankind's probable technological abilities so far in the future. As Arnold Boothroyd of the University of Toronto points out:

> By a thousand million years, if the human race survives its own inventions, it could be capable of doing whatever it wanted to survive. In principle there should be a way to shift Earth to a safer orbit farther from the Sun. A billion years is a lot of time. If we come up with something, it will probably be much sooner than that. Just think what we've done in the last few hundred years.[9]

However, a time will eventually come when it will no longer be possible to 'come up with something'. We live in a highly privileged epoch of the universe, a very special time called the Stelliferous Era, in which all its energy comes from the radiation of stars. It is not merely a good time to be alive, it is the *only* time to be alive – anywhere. We do not live in a central time any more than we live in a central place. Four centuries ago, as we have seen, Copernicus showed that the Sun, not the Earth, is the centre of the solar system; and in the twentieth century Edwin Hubble discovered that the Milky Way galaxy, hitherto thought to be the entirety of the cosmos, was only one in many billions of galaxies. At a stroke, it is said, Hubble enlarged the universe by a factor of many billions, and made it correspondingly more interesting.

In 1997 Fred Adams and Gregory Laughlin of the University of Michigan took a further step. What Copernicus and Hubble had done for our conceptions of space, they did for time. They showed that the Stelliferous Era is but a fleeting instant amidst immeasurable future ages.[10] 'We quickly run out of words to describe the time periods we're talking about,' said Adams. 'You're familiar with millions of years and billions of years and trillions of years, but we need to consider timescales that are very much longer than that.'[11]

Until recently, there was disagreement over whether the universe had any long-term future at all. Would it continue to expand for ever, or would it collapse in a 'big crunch' about 20 billion years hence, leaving no trace of itself at all? In the latter case, time and space would vanish, and no record of anything whatsoever would remain. The controversy centred on how much *matter* the universe contained. If there was a sufficient amount of it, then the present expansion would be halted by gravity, and the collapse would follow. But if not, then the universe would literally last for eternity. In 1998, three teams of astronomers announced independently that this was the case. The cosmos contained only a fifth of the amount of matter needed to bring about a big crunch.[12]

And so we are back to the immense timescales described by Adams and Laughlin. To depict them, they have invented the term 'cosmological decades'. This is a considerably more useful unit of time than an 'aeon' or 'eon', a vague and poetical word which merely means a vast age of undefined length. A cosmological decade, which I will shorten to 'decade', does not mean 10 years. It means a number of years expressed in powers of 10 – that is, 10 multiplied by itself any given number of times. Thus, dating from about 15 billion years ago when the universe was created, we live today in cosmological Decade 10, which is 10 multiplied by itself 10 times, 10 to the 10th power, or 10 billion years.* (In this kind of study, nobody cares overmuch about the difference between 10 billion and 15 billion years.)

Let us move forwards like Wells's Time Traveller. Two or three billion years hence, a new crisis starts in the core of the Sun. The mass of helium ash becomes steadily hotter and bigger as more and more hydrogen is burned. At length, about five billion years from now, these nuclear wastes reach the critical temperature of 90 million degrees, and the fatal Helium Flash occurs. The helium itself ignites in a nuclear explosion that, within a matter of hours, tears apart the Sun's outer layers. During the subsequent 30 million years, the helium atoms themselves fuse explosively to form atoms of oxygen, carbon, and neon. The Sun expands into what astronomers call a 'red giant'. Its diameter increases from just over a million kilometres to nearly 500 million. The entire solar system becomes uninhabitable. The Earth, if it is still in its present orbit and Whitehead's scheme has not been undertaken, is

* I will refrain here from using the notation 10^{10}, etc., since this might cause confusion with the reference numbers to the Notes at the end of the book.

swallowed up and vaporised. The only legacy of our world, with all Earth's man-made artefacts and monuments making a minuscule contribution, is a tiny increase in the metal content of the outer layers of the swollen Sun.[13]

The same fate will by this time have overtaken all the Sun-like stars that are now shining. The more massive ones will have exploded into supernovae or collapsed into black holes. Only the smallest stars, being cooler and converting their hydrogen into helium much more slowly – 'real misers', as Laughlin put it – will last many times longer. They may shine for another 10 trillion years, up to Decade 13.*

Meanwhile the hydrogen that escapes from dying stars and supernovae creates many generations of new stars, which in turn flare up and die, releasing fresh hydrogen to provide the nuclear fuel for still more stars. But this process cannot continue for ever. More than 90 per cent of the atoms in the universe are at present hydrogen, but since hydrogen will be constantly converted into heavier elements in the cores of stars, there will eventually be a shortage of hydrogen atoms. The universe will darken, increasingly filled with metal – the final stage of nuclear fusion – as an ever-greater part of its mass that has not collapsed into black holes and neutron stars consists of iron and nickel rather than hydrogen.

By Decade 15 (that is, 1 followed by 15 zeros), the formation of all Sun-like stars will have ceased. All that will remain to produce light will be the burned-out relics of once healthy stars, white dwarfs – stars the size of the Earth but as densely packed with matter as the Sun. This will be the Degenerate Era. This term has nothing to do with moral deterioration, but is used in its *mechanical* sense, meaning, as the *Oxford Dictionary* puts it, matter that has 'declined from a higher to a lower state'.[14] Or, as a physicist would put it, compressed matter whose atoms have been stripped of their electrons. The Degenerate Era will last from Decade 15 to Decade 30, an age-long epoch of darkness. In this period (in years of 1 followed by 37 zeros) there will be rare supernova explosions as two white dwarfs

---

* These cosmological decades date from the beginning of the universe, not from today. But as the decades get longer, the difference of 15 billion years becomes increasingly trifling. Fifteen billion is a fraction of 1 per cent of 10 trillion.

collide. This will happen about once every thousand trillion years in a galaxy the size of the Milky Way. Because what was once the starry sky will then be so dark, these bursts of energy will be incredibly spectacular, if only a human eye could be present to witness them.[15]

A slight qualification is needed here. It *would* in fact be technically possible for human eyes to witness these events. People could travel forwards in time for countless cosmological decades and observe this process of fatal decay. How? By travelling in a starship at a speed extremely close to the speed of light – much closer than the 92 per cent predicted in this book for routine flights. Perhaps as fast as the Oh-my-God particles. Recall that an extremely fast journey is a journey into the future. But it would be a melancholy expedition, for there would be no way of returning from it to the near present. It is hard to imagine that many people will volunteer.

Gradually, then, from about Decade 30, a new stage of the universe comes into being, the Black Hole Era, in which the white dwarfs and all the remaining debris from the long-gone Stelliferous Era merge and collapse into black holes. Black holes are normally thought of as collapsed stars, or collapsed clusters of stars, with such strong gravitational fields that nothing, not even light, can escape from them.*

Meanwhile, a still more fatal mechanism is at work, the destruction of atoms themselves. *Proton decay* it is called, the breakdown of the atomic nucleus. *All* atoms have broken down into elementary particles by about Decade 37.[16] I emphasise the word 'all', since even today some atoms are disappearing. A person who lives to the age of 100 will lose about one atom from his body during his lifetime.[17]

I must explain briefly why protons are going to decay. A generally accepted principle in physics, known as the 'grand unified theory', shows that three fundamental forces of nature –

---

* The physicist John A. Wheeler, who invented the term 'black hole', justified it as follows. Imagine a ballroom, where black-suited young men are dancing with young women in white. The lights are dimmed and you can only see the women. While the men in their black suits become invisible, *their existence can be deduced* from the movements of the women. The men are like black holes, and the brightly clad women are like normal stars whose orbits are being distorted by the holes.

gravity, the electromagnetic force and the nuclear force that holds atoms together – are merely different manifestations of the same force. Now this unification not only merges the three forces, it also joins together two classes of sub-atomic particles, leptons and quarks. Over sufficiently long timescales they become interchangeable, so that protons decay into leptons.[18] In short, the proton, the very bedrock of the atom and hence of all matter, becomes unstable.

Proton decay is the doom of life. Never again can planets or stars re-form, never again can new species of plants and animals come into being, under any imaginable arrangements of physics and biology. For not only will the absence of hydrogen atoms make star formation impossible; not only will the absence of carbon atoms make life as we know it impossible. The absence of *all* atoms makes it impossible for *any* matter as we know it to exist outside black holes.

By Decade 65, even black holes, which do not consist of complete atoms but only crushed, degenerate matter, will have started to disappear. As Stephen Hawking demonstrated in the 1970s to the astonishment of many scientists, black holes do not remain black for ever. They *leak* light. They evaporate, and all their matter is returned to the outside universe. The more massive the black hole, the longer it takes to evaporate, but in the end evaporate they must.[19] First, the black holes that were once single stars will have ceased to exist; and by Decade 98, the much more massive black holes that were once the cores of galaxies will have gone.

Now, at Decade 100, begins the final and everlasting phase, the Dark Era. There is nothing in the universe whatsoever except for stray particles. This will be the state of the cosmos *10,000 trillion trillion trillion trillion trillion trillion trillion trillion years* from now. As Adams put it: 'once the black holes have radiated away, the universe will consist of a diffuse sea of electrons, positrons, neutrinos and radiation'.[20]

Here is a summary of these cosmological decades:[21]

Event	*Decade*
Creation of the universe	0
First possible stars	6
Stelliferous Era begins	9
Formation of the solar system	9.5

The present epoch	10
Our Sun dies	10.2
Star formation ceases	14
Degenerate Era begins	15
Black Hole Era begins	30
All atoms have evaporated	37
Most massive black holes evaporate	98
Everlasting Dark Era begins	100

It could be easy to misunderstand this arithmetic. One could mistakenly view the length of each era as a percentage of the whole. One might glance at the above summary, note that the present epoch lasts from Decades 10 to 15, and imagine that this was 5 per cent of the lifetime of the universe. But to do this is to forget that the decades are measured, in years, by powers of 10, from the Creation. And so the period in which it will be possible for ships to fly from star to star is not 5 per cent of the period of time considered by Adams and Laughlin, but an incredibly tiny .000000000000000000000000000000000000000000000000000000 00000000000000000000000000000000000001 per cent of it. That is a decimal point followed by 85 zeros and then a 1. Even doing this sum with my pocket calculator I had to cheat, since the machine will not accept numbers so close to the infinitesimal.[22]

The Era of Starships, to use a new phrase, although it may seem long to human minds, will thus be an almost infinitesimal moment in actual time. Nothing can be done to prolong it. As we brood about this bleak and terrible future, it will eventually become apparent that we have a duty to ourselves to explore the stars and their millions of attendant planets while they are still there.

For with the Dark Era comes the end of ends. Black night will reign supreme for ever. It will be a cosmos of nothingness. Nor will conditions be any different in other universes. In them also atoms will have broken down.

Moreover the findings of Adams and Laughlin will be a shock to people with religious convictions. For if there is a supreme being who presides over the universe and all creatures who live in it, how can this purpose be served when there are no creatures? A god, it has always been supposed, exists to reward us for our virtues and punish us for our sins. But how can that be done when there is no one around to be virtuous or sinful? I fancy that

Giordano Bruno would have felt vindicated by the work of Adams and Laughlin, for it seems inconceivable that any rational system of philosophy could reconcile a deity with a dead universe.

There will be nothing to show that it was ever otherwise. Of man's dreams and ambitions and achievements no trace will remain. They will be recorded in no books and in no records, for without atoms there can be no books or records. To us, surrounded by the comforts of solid matter, the end of the universe must appear infinitely tragic and meaningless. But the Dark Era is far off. Between then and now there is much to be done.

Appendix I

Milestones to the Stars

Civilisation is a stream with banks. The stream is sometimes filled with blood from people killing, stealing, shouting, and doing things historians usually record. While on the banks, unnoticed, people build houses, make love, raise children, sing songs, write poetry and even whittle stones. The story of civilisation is what is happening on the banks. Historians are pessimists because they ignore the banks of the river.

Will and Ariel Durant, *The Story of Civilisation*

Every technological and scientific step undertaken now and in the future, no matter how seemingly trivial, is another mark on the road to practical interstellar flight. In fact, virtually the entire scientific-technological enterprise is already unwittingly contributing to the goal of starflight.

Eugene Mallove and Gregory Matloff,
The Starflight Handbook

To illustrate the important statement by Mallove and Matloff, here is a chronological list of the most significant breakthroughs in knowledge, know-how, and natural events in the past, present, and future that will lead us to the stars.

The two authors are no doubt right to say that *every* such breakthrough is another mark on the road, but to list them all would require a book as long, if not longer, than the 20-volume *McGraw-Hill Encyclopedia of Science and Technology*, and so I will desist. But those I cite here, although doubtless far from complete, should serve as a useful guide.* It seems particularly interesting that almost all of them – except those made for military purposes – have been the work of individuals rather than governments.

About 15 billion years ago. The creation of the universe.
About 10 billion years ago. The formation of our Milky Way galaxy.
About five billion years ago. Formation of the Sun and planets.
About 3,800 million years ago. The dawn of life on Earth.
About 65 million years ago. Probable impact of a giant asteroid that killed most of the dinosaurs, making possible the evolution of *Homo sapiens.*
About 500,000 years ago. The first use of fire.
About 40,000 years ago. Australian aborigines make the first formal attempt to classify the stars.
About 11,000 years ago. End of the last Ice Age, which made possible the rise of civilisation.
About 8,000 years ago. The invention of agriculture, which ends the epoch of hunter-gathering and starts the age of city states and empires.
About 3000 BC. The mastery of the horse, bringing about the attainment of Speed 4. The invention of the abacus, the first calculating device. The invention of writing.
About 2630 BC. Imhotep, the grand vizier of Egypt, builds the step pyramid at Saqqara in Memphis, the first massive man-made structure.
2250 BC. The first map. A clay tablet shows the river Euphrates flowing through Mesopotamia.
About 700 BC. The dawn of the Iron Age brings about great improvements to weapons and armour, facilitating the rapid mass movements of peoples.
312 BC. Appius Claudius, Censor of the Roman Republic, builds the first all-weather road, the Appian Way, from Rome to Brindisi.

---

* Some of them – but only some – were listed in the Appendix called 'The March of Knowledge' in my book *Eureka! and Other Stories.* I have now considerably revised and updated the list.

300 BC. Euclid, in his *Elements*, systematises the laws of arithmetics and geometry.

About 250 BC. Aristarchus of Samos calculates with reasonable accuracy the circumference of the Earth.

About AD 140. Claudius Ptolemy publishes the first comprehensive map of the northern naked-eye stars, known as the *Almagest*, or the 'greatest'.

About 250. Diophantus of Alexandria produces the first book on algebra.

285. Pappus of Alexandria describes five machines in general use: the cogwheel, the lever, the pulley, the screw, and the wedge.

372. The Huns introduce the stirrup into Europe, vastly improving the art of horsemanship.

595. An unknown Indian proposes the use of the zero in arithmetic.

815. An unknown Arab complements this achievement by inventing the modern system of numerals, making it easy to multiply by 10.

963. Al Sufi, in his *Book of the Fixed Stars*, refers to the 'nebulae', which centuries later are shown to be galaxies beyond our own.

About 1250. The invention of the compass.

1257–68. The friar Roger Bacon predicts the eventual coming of motor cars, motor boats, powered aircraft, and parachutes. But because he had offended the Church, his *Opus Major* is not published until 1733.

1271. Marco Polo begins his journeys to China.

1303. First recorded use of spectacles for reading.

1313. Gunpowder invented by Berthold Schwartz, a German friar.

1337. The first scientific weather forecast, made by William Merlee of Oxford.

1418. Prince Henry the Navigator of Portugal establishes his school for navigation.

1454. Johannes Gutenberg builds the first printing press, a technology that leads to the downfall of ecclesiastical power.

1492. Christopher Columbus reaches the Americas. This is followed by many other pioneering voyages of discovery, including those by the Cabots, father and son, Vasco da Gama, Ferdinand Magellan, and others.

1503–16. Leonardo da Vinci fills his notebooks with drawings of futuristic inventions.

1513. Nunez de Balboa discovers the Pacific Ocean.

1527. Paracelsus lectures on medicine at the University of Basle and later produces the first manual of surgery.*
1543. Nicholas Copernicus publishes his book *On Revolutions*, showing that the Earth revolves around the Sun.
1550. Rheticus (the adopted name of Georg Joachim von Lauchen) publishes trigonometric tables.
1550–90. Leonard Digges and his son Thomas invent the first telescope, which is probably kept a secret at the time by Elizabeth I's government for security reasons.†
1556. Georg Agricola publishes a study of mineralogy.
1572. Tycho Brahe makes astronomical observations that lead to Kepler's laws of planetary motion.
1595. Geradus Mercator publishes his global map.
1600. Giordano Bruno is burned at the stake by the Inquisition in Venice for suggesting that all stars have planets, provoking interest in the very ideas that his persecutors want to suppress.
1608. Johann Lippershey of Holland produces the first telescope that becomes widely available.
1609. Galileo looks at the night sky through one of Lippershey's telescopes. He sees the four largest moons of Jupiter, mountain ranges on the Moon, and many more stars than can be seen through the naked eye. Johannes Kepler publishes his laws of planetary motion.
1614. John Napier invents logarithms.
1620. Francis Bacon publishes his *Great Instauration of Learning*, which, along with his other great works of philosophy, analyses the scientific method and accelerates the growth of modern science.
1628. William Harvey describes the circulation of the blood.
1637. Pierre de Fermat scribbles in the margin of a book his famous Last Theorem, that '$X^N + Y^N$ never equals Z^N if N is a whole number greater than 2', adding: 'I have found a marvellous proof of this theorem, but the margin is too narrow to contain it.' It provokes more erroneous proofs than

---

* Paracelsus was the adopted name of Theophrastus Bombastus von Hohenheim. Aggressive and dogmatic in manner, he gave to the world the adjective 'bombastic'.

† A discovery made in 1991 by Colin Ronan, then president of the British Astronomical Association.

any other theorem in mathematics, and remains unproven until 1993.*

1654. Fermat and Blaise Pascal develop the theory of probability.
1660. Charles II founds the Royal Society of London.
1665. Giovanni Cassini determines the rotations of Jupiter, Mars, and Venus.
1671. Gottfried Leibniz insists upon the existence of 'ether' in space, which is not disproved until the Michelson–Morley experiment of 1887, which in turn paves the way for Einstein's special theory of relativity.
1675. Olaus Romer makes the first roughly accurate estimate of the speed of light.
1681. Engineers of Louis XIV start to build a pumping system to carry water from the Seine to the fountains of Versailles.
1687. Isaac Newton publishes his *Principia*, which contains his three Laws of Motion and his Law of Universal Gravitation.
1689. Father Richard, a Jesuit astronomer, discovers that Alpha Centauri – or Rigil Kent – is a double star system.
1705. Edmund Halley discovers the true nature of comets.
1742. Anders Celsius invents the Centigrade scale of temperatures.
1752. Benjamin Franklin invents the lightning conductor.
1761. Mikhail Lomonosov discovers that Venus has an atmosphere.
1766. Henry Cavendish identifies hydrogen, the most abundant element in the universe.
1772. Daniel Rutherford and Joseph Priestly independently discover nitrogen.
1777. Antoine Lavoisier, the father of modern chemistry, proves that air consists mostly of oxygen and nitrogen.
1781. William Herschel discovers the planet Uranus.
1783. The Montgolfier brothers make the first manned flight, in a balloon.
1801. Giuseppe Piazzi discovers Ceres, the largest asteroid. Joseph Lalande completes his catalogue of 47,390 stars. Robert Fulton,

---

* It is impossible to prove that achievements in pure mathematics like Fermat's Last Theorem will ever be of any practical use in interstellar travel. But one cannot be sure. After all, prime numbers, long thought interesting but useless, became from 1978 onwards an integral part of the world's most secure secret code. And so I include it.

inventor of the steamship, builds the first submarine.

1802. John Dalton introduces atomic theory into chemistry. William Herschel discovers that most stars are binaries.

1807. Jons Berzelius distinguishes the two great branches of chemistry, organic and inorganic. He later publishes the molecular weights of more than 2,000 chemical compounds.

1823. Charles Babbage starts his unsuccessful efforts to build a mechanical computer.

1831. Michael Faraday demonstrates an electric generator.

1832. Thomas Henderson determines the distance to Alpha Centauri as 4.3 light years.

1835. The journalist Richard Adam Locke hoaxes tens of thousands of readers of the *New York Sun* into believing that astronomers had found an alien civilisation on the Moon. Although exposed, the hoax creates a permanent popular interest in space travel and alien life.

1839. Louis Daguerre develops photography.

1846. Urbain Leverrier and J.C. Adams independently discover the planet Neptune.

1850. William Bond takes the first astronomical photograph, of the star Vega.

1859. Charles Darwin proposes the theory of evolution in *The Origin of Species*.

1864. Julian Sachs demonstrates photosynthesis.

1869. Dimitri Mendeleyev develops the periodic table of the chemical elements.

1870. Louis Pasteur and Robert Koch establish the germ theory of disease.

1871. Simon Ingersoll invents the pneumatic rock drill.

1873. James Clerk Maxwell discovers that electricity and magnetism are different manifestations of the same fundamental force.

1874. Lord Kelvin pronounces the Second Law of Thermodynamics.

1876. Alexander Graham Bell invents the telephone.

1880. Thomas Edison and Joseph Swann independently invent electric lighting. Friedrich Beilstein starts his vast classification of organic compounds.

1882. Albert Michelson measures the speed of light as 300,000 kilometres per second (1.1 billion k.p.h.)

1887. Michelson and Edward Morley conduct their famous

experiment showing that the ether wind does not exist (see *1671*), paving the way for Einstein's special theory of relativity.

1888. Heinrich Hertz discovers radio waves.

1895. Wilhelm Rontgen identifies X-rays.

1898. Marie Curie discovers radioactivity, in the elements polonium and radium.

1903. Wilbur and Orville Wright make the first powered flight.

1905. Einstein publishes his special theory of relativity, showing that no material object can reach the speed of light, but that time slows down at speeds close to that of light. Latent in it also is the concept of antimatter.

1906. R.A. Fessenden makes the first radio broadcast.

1909. Leo Baekerland invents Bakelite, the first of the plastics.

1911. Heike Kamerlingh-Onnes discovers superconductivity.

1913. Henry Norris Russell and Einar Hertzsprung plot the brightness and colours of stars in their Hertzsprung–Russell Diagram, which explains the evolution of stars from birth to death.

1914. Walter Adams and Arnold Kohlschutter determine the absolute luminosity of stars from their spectra alone, making it possible to discover the distances of millions of distant stars.

1915. Robert Innes finds Proxima Centauri, the closest star to the Sun.

1916. Einstein publishes his general theory of relativity, introducing the idea of gravity being caused by curved space-time. It indirectly predicts the existence of black holes. It also eventually gives birth to the fictional 'warp drive' of *Star Trek*.

1918. Harlow Shapley determines the shape of our Milky Way galaxy.

1919. Einstein's general theory is proved correct during an eclipse of the Sun. Lord Rutherford conducts the first experiment in nuclear fission.

1921. Hermann Oberth writes his dissertation *The Rocket into Interplanetary Space*.

1924. Edwin Hubble discovers the existence of billions of galaxies beyond the Milky Way. Arthur Eddington discovers that the brightness of a star is related to its mass.

1925. John Logie Baird demonstrates television.

1926. Robert Goddard fires a liquid-fuelled rocket. Hubble discovers that the universe is expanding.

1928. Frédéric and Irène Joliot-Curie create artificial radioactivity.

1930. Clyde Tombaugh discovers the planet Pluto. Paul Dirac predicts the existence of antimatter. Vannevar Bush builds the first general-purpose computer.
1932. Carl Anderson photographs an antiparticle.
1935. Karl Jansky discovers radio waves from outer space, starting the science of radio astronomy. Robert Watson-Watt develops radar. James Chadwick discovers the neutron.
1937. Frank Whittle builds the first jet engine.
1939. Igor Sikorsky builds the first helicopter. J. Robert Oppenheimer and Hartland Snyder predict the existence of black holes.
1942. Enrico Fermi achieves the first self-sustaining nuclear fission reactor. John von Neumann builds the first high-speed computer.* Invention of magnetic tape for storage of data.
1943. Jacques-Yves Cousteau demonstrates the aqualung.
1945. The first atomic bomb is detonated. Arthur C. Clarke proposes stationary satellites for communication.
1947. Charles Yeager breaks the sound barrier in an aircraft, attaining Speed Seven. Invention of the transistor.
1948. Von Neumann works out the mathematics of machines which could in theory make identical copies of themselves. The 200-inch (five-metre) Mount Palomar telescope goes into service.
1950. Claude Shannon builds a chess-playing machine, arguably the forerunner of all computer games. Alan Turing proposes the 'Turing Test' to test the intelligence of a computer against the human mind.
1951. The first electric power from atomic energy.
1952. The first hydrogen bomb exploded.
1953. James Watson and Francis Crick discover the molecular structure of DNA, the core of all life. It is found to be essentially the same as the sequence for machine reproduction proposed by von Neumann in 1948, suggesting that machines can one day clone themselves.
1957. Russia's *Sputnik 1*, the first spacecraft, launched into orbit, achieving Speed 9.

---

* But thousands of times slower and thousands of times bigger than a modern personal computer!

1958. The US nuclear submarine *Nautilus* sails under the North Polar icecap.

1960. Robert Bussard proposes his interstellar ramjet. The weather satellite *Tiros I* transmits television pictures of cloud cover round the world.

1961. Yuri Gagarin is the first human in space. President John Kennedy commits the United States to landing a man on the Moon before 1970.

1963. Maarten Schmidt discovers the brightest and first known quasar, 3C 273.

1965. Edward White makes the first space walk.

1967. Jack Kilby of Texas Instruments designs the first pocket calculator.

1969–72. Six manned landings on the Moon. Important lunar resources are discovered, including the valuable isotope helium-3, which is very rare on Earth. The achievement also transforms terrestrial technology, in particular microelectronics.

1970. The safe return to Earth of three astronauts in *Apollo 13*, after their spaceship had been crippled by an explosion, shows that space need not be hostile. President Richard Nixon orders the construction of space shuttles. Donald Keck and Robert Maurer develop optical fibres for communication.

1973–4. America's *Skylab*, the first space station, manned by three teams of astronauts. For the next 12 years, Russia launches four *Salyut* space stations. Its most successful station, *Mir*, is launched in 1986.

1974. Alan Bond proposes his ram-augmented interstellar rocket.

About 1975. Shizuo Takano invents television video recorders.

1976. The *Viking* spacecraft soft-lands on Mars.

1977. Chuck Peddle and Stephen Wozniac respectively build the first successful personal computers, the Commodore PET and the Apple II.*

1978. Ronald Rivest, Adi Shamir, and Len Adelman invent the RSA cipher, an easy-to-use and almost impenetrable system of

---

* An example of the tremendous advances in computing power since this time is given in Jenkins, *The Number File*, in which he states that an Apple II would take 200 years to find the prime factors of the 24-digit number 984,073,151,544,289,893,895,597. In 1998 my Pentium I computer did this sum in less than 10 minutes.

secret writing for electronic communication, based on the intractability of finding two large prime numbers hidden in a much larger number.

1981. The first shuttle flight into space.

1983. President Ronald Reagan orders the construction of a manned space station much larger than *Skylab*. Tsutomu Ishii writes the names of 184 countries on a single grain of rice and the words 'Tokyo, Japan' on a human hair.

1986. Europe's *Giotto* spacecraft penetrates Halley's Comet.

1987. The supernova explosion '1987A', in the Greater Magellanic Cloud, emits traces of iron and nickel, proving that the heavier elements in our bodies came from exploding stars.

1988. Discovery of a pulsar star that rotates like a clock which only gains or loses a second in three million years.

1989. The spacecraft *Voyager 2* flies by Neptune, 4.3 billion kilometres from Earth, discovering eight moons and three rings. CERN's Large Electric Positron Collider, a particle accelerator with a circumference of 27 kilometres, comes into operation.

1990. The Hubble Space Telescope is launched into orbit. So are the ultraviolet observatories ROSAT and Astro-1.

1991. The *Galileo* space probe flies past the asteroid Gaspra, approaching it to within 26,000 kilometres. The first experimental controlled production of nuclear fusion energy, at the Joint European Torus, at Culham, Oxfordshire.

1992. George Smoot, using the Cosmic Background Observer satellite (COBE), detects the ripples of the Big Bang that started the universe some 15 billion years ago. Sanyo Electric, in Japan, makes transistors from superconducting ceramics. They are 10 times faster than those made from semiconductors.

1993. The demonstration flight of a 'space clipper', which would require a ground staff of only three compared with 20,000 in the case of the space shuttles, to take people and cargo into orbit for a price of $2,700, the approximate economy-class round-trip fare from Britain to Australia. Andrew Wiles proves Fermat's Last Theorem of 1637, one of the most baffling problems in pure mathematics.

1994. Alexander Wolcszcan detects three Earth-sized planets orbiting a pulsar – which for that reason would be uninhabitable. Twenty-two fragments of Comet Shoemaker-Levy 9 hit Jupiter, showing the violence of the impact on a planet of a comet or an asteroid. It also shows that the presence of a gas giant in a stellar system can be a long-term safety factor, minimising the number of such impacts on an Earth-sized planet.

1995. Michel Mayor and Didier Queloz discover a planet orbiting the Sun-like star 51 Pegasi, 42 light years from Earth. In the next two years a dozen other planets are found circling Sun-like stars. But because of the limits of telescope technology, these planets are all giant, uninhabitable worlds, gas giants more massive than Jupiter. William McLaughlin suggests that the ancient Doctrine of the Holy Trinity, in which three entities – or in this case microchips – deciding by majority vote when in disagreement could represent the future of 'intelligent' computer architecture. Dean Pomerleau rides in a driverless car from Pittsburgh to Washington.

1996. NASA scientists, in a meteorite discovered in the Antarctic, claim to have found evidence that life once existed on Mars. The claim remains controversial, but excites great interest in Martian exploration.

1997. NASA's Pathfinder robot roams the surface of Mars. Fred Adams and Greg Laughlin predict that the universe will be lifeless for more than 99 per cent of its future because there will no longer be any atoms.

1998. The *Prospector* spacecraft detects between 300 million and a billion tons of water in the form of ice in the polar regions of the Moon, a discovery that promises to facilitate space travel and colonisation.

1999. James Benson, chief executive of the private company SpaceDev, claims the ice-bearing asteroid Nereus as the company's private property.

And what of the future? The only four events and their dates we can predict with certainty are:

2026. The asteroid 1997 XF11, more than a kilometre in diameter, will pass within 960,000 kilometres of Earth, providing an excellent opportunity for a manned landing and an examination of its materials.

About 32,000 years from now. The spacecraft *Pioneer 11*, unmanned and with its instruments long dead, will reach the vicinity of the star Ross 248, 10.3 light years from Earth. This star is one of those unlikely to have habitable planets. But by such a time our descendants will probably have learned how to build worlds from the most unpromising materials, and so *Pioneer 11* may be recaptured at the end of its age-long voyage and placed in a museum.

About 1,500 million years from now. The Sun will become so hot that the Earth will become uninhabitable. At that time, if any humans are left on the planet, starships will become a necessity of life.

About 1,000 trillion years from now. All the stars, even those yet unborn, will have burned out. There will be no more habitable planets, and starships will become pointless.

Appendix II

'Beacon Stars' for Navigation

In Chapter 12, 'The Star on the Starboard Beam', I mentioned the necessity of having three 'beacon stars' by which the ship could calculate its position.

The requirements were that two of them must be very bright and very far away, while the third must be comparatively close by. In the early stages of the voyage, when the ship was travelling much slower than light and relativistic effects had not yet taken effect, the crew would calculate their position by measuring the changing angles between the nearby star and the two much more distant ones.

They would also be useful when the ship was decelerating, approaching its destination star, and the sky, no longer distorted by relativistic effects, was resuming its normal appearance.

For navigational aid when the ship is travelling at a substantial percentage of the speed of light, see Appendix III.

Here is a list of suitable 'navigational stars', all of which are naked-eye objects and were used by terrestrial mariners before the days of modern navigational instruments.* They are given in alphabetical order. (I have also listed the Pleiades, although this is not a single star but a bright, tightly knit cluster of stars.) The crew could choose from any of these, depending on their direction. The nearby stars are marked in bold, the rest in normal type. The scientific name of each star will indicate its position by

---

* They are taken from lists in Michael E. Bakich, *The Cambridge Guide to the Constellations*.

showing its constellation, and I give its brightness (apparent magnitude) as seen from Earth.

Star name	*Scientific name*	*Distance (light years)*	*Brightness*
Acamar	Theta Eridani	55	2.06
Achenar	Alpha Eridani	85	0.46
Acrux	Alpha Crucis	270	0.79
Adhara	Epsilon Canis Majoris	620	1.50
Aldebaran	Alpha Tauri	65	0.85
Alioth	Epsilon Ursae Majoris	70	1.77
Alkaid	Eta Ursae Majoris	210	1.86
Alnair	Alpha Gruis	65	1.74
Anlilam	Eta Orionis	1,600	1.70
Alphard	Alphae Hydrae	95	1.98
Alphecca	Alpha Coronae Borealis	72	2.23
Alpheratz	Alpha Andromedae	120	2.06
Alsuhail	Lambda Velorum	148	2.21
Altair	**Alpha Aquilae**	**17**	**0.77**
Ankaa	Alpha Phoenicis	83	2.39
Antares	Alpha Sorpii	400	0.96
Atria	Alpha Triangulum Australis	80	1.92
Avior	Epsilon Carinae	340	1.86
Bellatrix	Gamma Orionis	450	1.64
Betelgeuse	Alpha Orionis	650	0.50
Canopus	Alpha Carinae	300	–0.72
Deneb	Alpha Cygni	1,500	1.25
Diphda*	Beta Cetis	60	2.04
Dubhe	Alpha Ursa Majoris	105	1.79
El Nath	Beta Tauri	270	1.65
Eltanin	Gamma Draconis	130	2.23
Enif	Epsilon Pegasi	543	2.39
Gacrux	Gamma Crucis	220	1.63
Gienah	Epsilon Cygni	57	2.46
Hadar	Beta Centauris	300	0.61
Hamal	Alpha Arietis	75	2.00
Kaus†	Epsilon Sagittarii	125	1.85

*Also sometimes called Deneb Kaitos.

† Also sometimes called Kaus Australis.

Kochab	Beta Ursae Minoris	100	2.08
Markab	Alpha Pegasi	86	2.49
Menkar	Alpha Ceti	362	2.53
Menkent	Theta Centauri	50	2.06
Miaplacidus	Beta Carinae	85	1.68
Mirfak	Alpha Perseii	570	1.79
Nunki	Sigma Sagittarii	250	2.02
Peacock	Alpha Pavonis	310	1.94
Pleiades	Taurus	410	2.00
Polaris	Alpha Ursae Minoris	360	2.02
Procyon	**Alpha Canis Minoris**	**11**	**0.38**
Rasalhague	Alpha Ophiuchi	49	2.08
Regulus	Alpha Leonis	85	1.35
Rigel	Beta Orionis	850	0.12
Rigil Kent	**Alpha Centauri**	**4.3**	**–0.27**
Sabik	Eta Ophiuchi	63	2.43
Schedar	Alpha Cassiopeiae	204	2.23
Shaula	Lambda Scorpii	300	1.63
Sirius	**Alpha Canis Majoris**	**8.7**	**–1.46**
Spica	Alpha Virginis	220	0.97
Zuben	Alpha Librae	56	2.75

Appendix III

Navigating a Starship at Close to the Speed of Light

Here is a simple BASIC program, by Keith Malcolm, to calculate the speed of a starship from observed stellar displacement. It measures the position of a star that normally lies at a right angle to the ship's direction of motion. It will of course work only when the ship is travelling at a significant percentage of the speed of light.

```
10 THETADEG=90:REM***set star angle when stationary to
   90 degrees
20 PRINT TAB(20)"Displacement of Speed
30 PRINT TAB(20)"guide star (psol)
40 PRINT TAB(20)"(degrees)"
50 PRINT
60 FOR LIGHTSP%=10 TO 90 STEP 10:
   VOVERC=LIGHTSP%/100
70 GOSUB 180:REM***routine to calc THETA1 from known
   theta and speed
80 THETA1$=STR$(FIX((THETA1+.005)*100)):REM***2 dps
   accuracy
90 PRINT TAB(20) LEFT$ (THETA1$, LEN (THETA1$)-2) "."
   RIGHT$ (THETA1$,2),,;LIGHTSP%
100 NEXT LIGHTSP%
110 FOR LIGHTSP%=91 TO 99 : VOVERC=LIGHTSP%/100
120 GOSUB 180
130 THETA1$=STR$(FIX((THETA1+.005)*100)):REM***2
   dps accuracy
```

```
140 PRINT TAB(20) LEFT$ (THETA1$, LEN (THETA1$)-2)
  "." RIGHT$ (THETA1$,2),,,;LIGHTSP%
150 NEXT LIGHTSP%
160 END
170 REM
180 REM****subroutine to calculate theta1 from known theta
  and voverc
190 THETARAD=THETADEG*.0174533:REM***convert to
  radians
200 COSTHETA1RAD=(COS(THETARAD)+VOVERC)/
  (1+VOVER- C*COS (THETARAD))
210 X=COSTHETA1RAD: GOSUB 240: THETA1 =
  ARCCOSX* 57.29578: REM***conv to degrees
220 RETURN
230 REM
240 REM****subroutine to return the arc cosine of x in
  arccosx
250 ARCCOSX=-ATN(X/SQR(-X*X+1))+1.5708
260 RETURN
```

Of the variables used in the program, LIGHTSP% is the percentage of the speed of light that the calculated angle of the guide star is shown for.

VOVERC is the ship's velocity divided by the speed of light (v over c, or v/c).

THETADEG (theta degrees) is the angle of the guide star when the ship is moving at negligible speeds. It is set at 90 degrees and does not change.

Using a formula from p. 181 of Eugene Mallove and Gregory Matloff's *The Starflight Handbook*, the subroutine in lines 180–220 calculates THETA1, the observed angle of the guide star, as the ship accelerates.

Glossary

Absolute zero. At –273, the coldest possible temperature in the universe, at which atoms cease to move. Nothing, as far as we know, is actually this cold, but temperatures a few fractions of a degree above absolute zero have been produced in laboratories.

Antimatter. Matter with the opposite electric charge to ordinary matter, and which explodes violently on contact with it, converting all its mass to energy. It is seen as a means of propelling starships.

Apparent magnitude. The brightness of a celestial object as seen from Earth. The brighter the object, the lower the object. The Sun, the brightest object in the sky, has an apparent magnitude of about –27, and the faintest stars that can be seen with telescopes about +20. With the naked eye one cannot usually see stars fainter than about +7.

Artificial gravity. If a sufficiently large spacecraft is rotated, gravity of any desired strength can be created inside it by centrifugal force. Even a small spaceship can be given artificial gravity. A long tether could be attached to it whose other end is fastened to an object with approximately the same mass as the ship. When the whole system is made to rotate, people in the ship would no longer be weightless.

Asteroid belt. A huge swath of asteroids, or minor planets, that lies between the orbits of Mars and Jupiter. A starship would probably be flight-tested above this region and safely clear of it.

Astronomical Unit (AU). The distance between the Earth and the Sun, 150 million kilometres, a standard measure of distance in the solar system.

Atmosphere. The mass of air that surrounds the Earth – or any other planet that is habitable or has been made so. It extends upwards (on Earth) to a height of about 200 kilometres.

BASIC. The oldest, simplest and best-known computer programming language, short for 'beginners all-purpose symbolic instruction code'. The most familiar version is IBM BASIC.

Beacon stars. Three very bright and very distant stars that would measure a starship's position by triangulation when it was moving at low speeds. Not to be confused with *guide star*.

Billion. A thousand million.

Black hole. An object in space with such a strong gravitational field that nothing, not even light, can escape from it.

Brightness of stars. See *apparent magnitude*.

Brown dwarf. A small star that is only 10 to 100 times more massive than Jupiter, just massive enough to shine with its own nuclear fusion. A brown dwarf too close to a star can use up all the material from which an Earth-sized planet might otherwise have formed.

Carbonaceous material. Any material, particularly a meteorite or an asteroid, that contains carbon.

Catalogue. Of the stars in our galaxy, only 563 have actual names. Beyond this, there are about 2,000 with Greek letters. See also *constellation*. The vast majority of stars appear in the Henry Draper Catalogue, last updated in 1930, with 1,072,000 stars. On a star chart, therefore, a faint star might be named HD (after Henry Draper) and then a number.

Constellation. Any one of the 88 regions into which the night sky is arbitrarily divided. Usually named after some mythological character, they are convenient as different directions in which to look for a particular star. The brightest star in a particular constellation is called Alpha, followed by the constellation name. The 24th brightest in that constellation is Omega. Thus Vega, the brightest star in the constellation of Lyra, is also called Alpha Lyrae. See also *catalogue*.

Cosmological decade. A phrase to denote an extremely long period of time, when such words as billions or trillions of years

lose their meaning. Cosmological Decade 30, for example, means a number of years of a 1 followed by 30 zeros.
Cryogenic materials. Materials that must be stored or transported at sub-freezing temperatures.
Cryonic suspension. The art of suspending life at extremely low temperatures from which it can later be revived.

Download, to. To extract information from a computer.

Element. One of the 109 fundamental substances of which the universe is composed. The natural elements (as distinct from man-made ones) range from hydrogen, the lightest, to the heaviest, uranium.
Escape velocity. The speed which a vehicle needs to reach to escape from the surface of a given planet and reach orbit.

Fission, nuclear. The energy created by the splitting of atoms. All atomic power stations at present operate by fission.
Fusion, thermonuclear. The process by which energy is created inside stars, in which elements are converted or 'fused' into other elements at gigantic temperatures, and which we hope to imitate, both in terrestrial and interplanetary power stations and in starship rocket engines.

G. The force of gravity on a world, or the force experienced in a spaceship due to its acceleration or rotation. The Earth's gravity is defined as 1 g.
Galaxy. A vast collection of stars, numbering tens or hundreds of billions. The Earth is near the edge of the Milky Way galaxy, which numbers some 250 billion stars. There are believed to be about 100 billion other galaxies in the universe. The term 'the galaxy' should be taken always to mean our Milky Way.
Geostationary orbit. An orbit where an object will stay always at the same point above the equator, invaluable for communication satellites. On Earth, this orbit is at an altitude of 36,000 kilometres.
Globular clusters. Tightly packed spherical clusters of stars containing 10,000 to millions of stars. Their stars are the oldest in the universe, and are older even than most galaxies.
Grand unified theory. A prediction that gravity, the electromagnetic force and the nuclear force that binds atoms together are

all different manifestations of the same fundamental force.

Gross human product. The total amount of wealth owned by mankind in its entirety, wherever it may be scattered in the universe.

Gross world product. The total wealth of a single planet such as the Earth. At present, this sum has been estimated at about $15 trillion. See also *wealth*.

Guide star. Star that lies at a right angle to a starship's direction of motion and which would be used as a speedometer when the ship reached relativistic speeds, by measuring the change of its angle. Not to be confused with *beacon stars*.

Hardware. Solid machinery, as opposed to 'software'.

Hertzsprung–Russell Diagram. A chart in which the true brightness of stars is plotted against their temperatures or colours, and which indicates also their masses and ages. Compiled in 1913 by Einar Hertzsprung and Henry Norris Russell.

Hybrid rocket engine. One with both oxidiser and unconventional propellant like polybutadiene, or synthetic rubber.

Hydrazine. A rocket propellant made of a single fuel that does not require an oxidiser. Used by a spacecraft to make mid-course corrections.

IBM BASIC. See *BASIC*.

IGS. Short for inertial guidance system. Measures a starship's rate of acceleration and changes of course, and the voyage's elapsed time, in order to calculate position and speed.

Inner solar system. The solar system within the orbit of Mars, and containing the inner planets Mars, Earth, Venus, and Mercury.

Interest, simple and compound. Two ways of calculating the periodic increase in a quantity, typically a sum of money. With simple interest, the periodic increase is always a percentage of the original amount. But compound interest is a percentage of the quantity *as increased in the last period*. Compound interest thus increases value much more rapidly than simple interest.

Interplanetary space. The space between the planets.

Interstellar space. The space between the stars.

Joule. A unit of energy named after the nineteenth-century physicist James Joule. A joule is approximately the amount of energy a person needs to climb one step of a staircase.

Kinetic energy. The energy released by the violence of a collision, equal to the mass of the impacting object multiplied by half the square of its speed.
K.p.h. An abbreviation for kilometres per hour.

Light, speed of. 1.1 billion k.p.h., or 300,000 kilometres per second. In mathematical equations this speed is written as c.
Light year. The distance which light travels in one year, 9.5 trillion kilometres, a standard measure of distance in the universe.
Little Green Men. A shorthand phrase for alien civilisations elsewhere in the galaxy – if they exist.

Milligram. A thousandth of a gram. There are thus a billion (10^9) milligrams to a ton.
Multiple star. What appears to be a single star but is in fact two or more stars orbiting one another. Alpha Centauri (or Rigil Kent) is a triple star system.

Nanotechnology. The art of making and using extremely small machines, based on the word 'nano', meaning a billionth.
NASA. The National Aeronautics and Space Administration, America's space agency.
Non-cryogenic materials. Materials that can be stored or transported at room temperature.

Oort Cloud. A vast cloud of some 100 billion comets that surrounds the solar system at a distance of about three trillion kilometres, 20,000 astronomical units, named after its discoverer Jan Oort.
Orbit. The path through space of one object round another.
Outer solar system. The solar system beyond the orbit of Mars, containing the 'outer planets' Jupiter, Saturn, Uranus, Neptune, and Pluto.

Parking orbit. The universe offers an unlimited number of possible orbits round different celestial bodies, many of them virtually everlasting. A parking orbit would be one where it was convenient to place a station for broadcasting, receiving and storing data.

Plasma. The fourth state of matter, beyond solid, liquid, and gas. It comes into existence when matter has been sufficiently heated, stripping its electrons from its atoms. Plasma has the advantage of responding to a magnetic field, so that in a ramjet starship engine, proposed by Robert Bussard, hydrogen in space could be collected by a 'magnetic scoop' and used as fuel.

Primates. The highest order of mammals, including man, apes, and monkeys.

Prime number. A number like 2, 3, 5, etc., that is only divisible by itself and 1.

Program. A set of instructions – or 'software' – which tells a computer what to do. A word-processing program is one of the best-known examples.

Pulsar, or **neutron star.** A super-dense star about the size of the Earth with the mass of the Sun. A cubic centimetre of it would weigh about a billion tons. It could not harbour a habitable planet because it would emit a constant stream of lethal X-rays and gamma rays.

Quantum computer. A machine that would operate in many different states, or, as some theorists believe, many different universes, simultaneously. By doing this, it would work billions of times faster than the fastest supercomputer.

RAIR. Short for ram-augmented interstellar rocket, a hybrid starship engine proposed by Alan Bond, which would combine both a ramjet and a nuclear fusion engine.

Ramjet. A jet or rocket engine which works by sucking in gas from outside with which it supplements its fuel.

Relativistic speeds. Speeds of a significant percentage of the speed of light, 50 per cent or more, at which phenomena predicted by Einstein's 1905 theory become important.

Relativity, special theory of. Formulated by Einstein in 1905, it shows – in particular – that no material object can exceed the speed of light, and that astronauts in a ship whose speed approached that of light would age more slowly than people on Earth.

Retro-rocket. A rocket that slows down a spacecraft, enabling it to enter atmosphere or descend to a planet.

Rigil Kent. A popular name for the Alpha Centauri star system, the closest stars to the Sun.

Robot. A machine controlled by electronic or other means that is programmed to perform tasks. I have used the term in this book to mean a movable machine, as opposed to a computer, which is usually stationary.

SETI. Short for the Search for Extraterrestrial Intelligence, a radio search for alien civilisations in space. Formerly run by NASA, it was cancelled by Congress in 1993 but revived by private contributors.

Software. Synonymous with *program.*

Solar power satellite. A large reflector in geostationary orbit which would beam down solar energy in the form of microwaves that would be converted into electricity.

Solar system. The Sun and planets and all the space between them. See also *inner solar system* and *outer solar system.*

Solar wind. A stream of high-speed particles that pours out from the Sun – or from any normal star – which will make possible the use of 'solar sails'.

Space. The entirety of the universe except for the Earth and its atmosphere.

Spiral arm. One of the great limbs of gas and debris that extend from a spiral galaxy like ours, a possible hazard to any inhabited star system inside one.

Stellar system. A system of one or more planets circling another star.

Stelliferous Era. The comparatively brief and fleeting present period of the universe, when all natural energy comes from the radiation of stars.

Supercomputer. A machine that operates millions of times faster than a personal computer.

Technological growth. The expansion of technical ability, a process that seems likely to continue indefinitely.

Terraform. To change a planet with a hostile environment into one similar to Earth's, or at least one that people can live on.

Time dilation. The shrinkage of time in a fast-moving ship, predicted by Einstein's 1905 special theory of relativity, and proved by many experiments to be correct.

Trillion. A million million, 1 followed by 12 noughts.

Virtual world, or **virtual adventure.** An imaginary world, created by computer software.

Wealth. The totality of resources available to a community. This includes such unrealised assets as education and all kinds of knowledge.

References and Notes

Introduction

1. The full text of Copernicus's offending passage may be found in my book *Eureka and Other Stories: A Book of Scientific Anecdotes*.
2. Giordano Bruno, *De l'Infinito, Universo, e Mondi* ('The World, the Universe and Infinity'), 1584.
3. Quoted in a brief biography of Bruno by John J. Kessler, posted on the Internet.
4. I have taken these facts about Bruno from Croswell's *Planet Quest*; the 1911 *Encyclopedia Britannica*, Vol. 4, pp. 686–7; and Kessler, op. cit.
5. From Singer, *Giordano Bruno: His Life and Thought*, pp. 175–7. Quoted by Croswell, op. cit., pp. ix–x.
6. Carlyle and Bismarck, quoted by Diamond, *Guns, Germs and Steel*, p. 420.

Chapter 1: AN ALIEN WORLD

1. For an overview of the exotic, super-strong, super-lightweight materials that will be available in the next century, see Ball's *Made to Measure*.

   A sign of the continuous progress being made in materials design is a laminate called 'Glare' of which the hull of the proposed Airbus A3XX superjumbo jet aircraft, with seats for 650 passengers, is likely to be made. Consisting of glass fibre tape pressed at high temperature between layers of aluminium, it is said to be 15 per cent lighter than conventional aluminium alloys, more resistant to fire, and ruggedly resistant to cracking. See Charles Goldsmith, 'For Airbus A3XX, a Weighty Decision is Unbearably Light', *The Wall Street Journal Europe*, 7 September 1998.
2. I have written the following short BASIC program to show how the

brightness of a star dictates what distance a planet should be from it to be habitable. It also shows how the mass of a star – compared with the Sun's – affects its lifetime. The rule is that the greater the star's mass, the shorter its lifetime, and vice versa.

This program is based on calculations in Gillett's book *World-Building*, but the responsibility for any errors it contains are my own.

The variables are MSLT$, meaning 'main sequence lifetime', the approximate number of years that the star will shine before entering its death-throes, and AU, the distance between the Earth and the Sun (150 million kilometres), a useful measurement of distance in astronomy. AM is the apparent magnitude of the star, as seen from Earth, and ABSM is the comparative brightness of the star if all stars were seen from the same distance of 10 parsecs or 32.6 light years. LUM, or luminosity, is the total amount of radiation that the star emits.

The symbols <= means 'is equal to or less than', and > means 'is more than'. The symbol < by itself merely means 'is less than'. MASS is the mass of the star compared with the Sun's mass that is arbitrarily given as 1.

From the known distance of the star and its apparent magnitude, the program deduces its absolute magnitude, its luminosity, and from the long subroutine that goes from Lines 20 to 270, its main sequence lifetime. A sample run is given at the end. Here is the program:

```
10 GOTO 280
20 REM * * * * * STAR MASSES TO LIFE TIME
  RATIOS * * * * * *
30 IF MASS<36 AND MASS>18 THEN MSLT$="four
  hundred thousand years"
40 IF MASS<=18.2 AND MASS>5.8 THEN
  MSLT$="about 3 million years"
50 IF MASS<=5.8 AND MASS>3 THEN MSLT$="about 7
  million years"
60 IF MASS<=3 AND MASS>2.6 THEN MSLT$="about
  470 million years"
70 IF MASS<=2.6 AND MASS>2.16 THEN
  MSLT$="about 720 million years"
80 IF MASS<=2.16 AND MASS>2.01 THEN
  MSLT$="about 1.2 billion years"
90 IF MASS<=2.01 AND MASS>1.75 THEN
  MSLT$="about 1.4 billion years"
```

```
100 IF MASS<=1.75 AND MASS>1.57 THEN
  MSLT$="about 2.1 billion years"
110 IF MASS<=1.57 AND MASS>1.43 THEN
  MSLT$="about 2.8 billion years"
120 IF MASS<=1.43 AND MASS>1.27 THEN
  MSLT$="about 3.7 billion years"
130 IF MASS<=1.27 AND MASS>1.09 THEN
  MSLT$="about 7.8 billion years"
140 IF MASS<=1.09 AND MASS>1.02 THEN
  MSLT$="about 9.5 billion years"
150 IF MASS<=1.02 AND MASS>.95 THEN
  MSLT$="about 12 billion years"
160 IF MASS<=.95 AND MASS>.88 THEN
  MSLT$="about 14 billion years"
170 IF MASS<=.88 AND MASS>.79 THEN
  MSLT$="about 19 billion years"
180 IF MASS<=.79 AND MASS>.72 THEN
  MSLT$="about 25 billion years"
190 IF MASS<=.72 AND MASS>.64 THEN
  MSLT$="about 35 billion years"
200 IF MASS<=.64 AND MASS>.55 THEN
  MSLT$="about 52 billion years"
210 IF MASS<=.55 AND MASS>.48 THEN
  MSLT$="about 78 billion years"
220 IF MASS<=.48 AND MASS>.42 THEN
  MSLT$="about 110 billion years"
230 IF MASS<=.42 AND MASS>.38 THEN
  MSLT$="about 150 billion years"
240 IF MASS<=.38 AND MASS>.32 THEN
  MSLT$="about 250 billion years"
250 IF MASS<=.32 AND MASS>.18 THEN
  MSLT$="about 1.2 trillion years"
260 IF MASS<=.18 THEN MSLT$="trillions of years"
270 RETURN
280 CLS:KEY OFF:PRINT"CONSTRUCTING
  HABITABLE ALIEN WORLDS"
290 INPUT"Distance of the star in light-years";LY$
300 LY=VAL(LY$):IF LY<=0 THEN BEEP:GOTO 290
310 PC=LY/3.26:REM CONVERT LIGHT-YEARS TO
  PARSECS
320 INPUT"Apparent magnitude of the star";AM$
```

```
330 AM=VAL(AM$):IF AM=0 THEN BEEP:GOTO 320
340 ABSM=AM+5-(5*LOG(PC)*.4343):REM
  CONVERTING TO BASE-10 LOGS
350 LUM=2.52∧(4.85-ABSM)
360 PRINT
370 PRINT"Its absolute magnitude is"ABSM"
380 PRINT"and its luminosity is";LUM"times the Sun's."
390 MASS=LUM∧.2632:PRINT"Its mass will therefore
  be"MASS"times the Sun's,"
400 GOSUB 20:PRINT"giving it a main sequence lifetime
  of"MSLT$
410 AU=SQR(LUM)
420 PRINT"To get the same luminosity of light that Earth
  gets,"
430 PRINT"the planet should be at
  about"INT(AU*150)"million kilometres from the star"
```

For a sample run, we can take the Sun-like star Tau Ceti, which is 11.41 light years away with an apparent magnitude of 3.49. The program shows that to get the same intensity of light and warmth that the Earth gets from the Sun, a hypothetical planet should be at about 65 per cent of the Earth's distance from the Sun, or 98 million kilometres. (Somewhat cumbrously, the program gives the absolute magnitude of Tau Ceti as 5.769625, its luminosity as .44274272 times the Sun's, and its mass as .799545 solar masses. Ideally, these numbers should appear respectively as 5.77, .442 and .78. But unfortunately I know of no way to round off numbers to two decimal places in BASIC.)

3. Alexander G. Smith, 'Settlers and Metals – Industrial Supplies in a Barren Planetary System', *Journal of the British Interplanetary Society*, Vol. 35, 1982, pp. 209–17.
4. Adrian Berry, ' "Oceans of ice" could support life on Mars', *Daily Telegraph*, 14 May 1990.
5. Roger Highfield, 'Water Finds Raise Hope of Alien Life', *Daily Telegraph*, 8 April 1998.
6. Gillett, op. cit., pp. 143–4.
7. There is a possibility that our Sun *does* have a companion star, with a highly elliptical orbit. Theorists have named this object, which would be almost invisible if it existed, Nemesis, the Death Star, since, when it approaches the Sun every 26 million years, it perturbs comets and increases the frequency of their collisions with Earth. Although its presence would account for the frequency of cometary collisions, no one has yet found it.

8. Gillett, op. cit., Ch. 7.
9. This number is considered ideal for planetary colonists. See a proposal for colonising Mars by Gary A. Allen (who gives the inexplicably precise number of 940), 'An Interplanetary Transportation System for Delivering Large Groups of People to Mars', *Journal of the British Interplanetary Society*, Vol. 48, 1995, pp. 373–86.
10. William Hodges, 'The Division of Labour and Interstellar Migration', a contribution to *Interstellar Migration and the Human Experience*, edited by Ben R. Finney and Eric M. Jones.
11. Ibid.
12. William Manchester, *A World Lit Only by Fire*, pp. 21–2.
13. The latest explanation of the Lost Colony of Roanoke Island is that they were faced with the worst drought for eight centuries, a disaster that even the most meticulous planning could not have prepared them for. See Aisling Irwin, ' "First American Colony Wiped out by Drought" ', *Daily Telegraph*, 24 April 1988.
14. John Geiger and Owen Beattie, *Dead Silence: The Greatest Mystery in Arctic Discovery*, Bloomsbury, London, 1993.
15. *Dictionary of National Biography*, Vol. 7, pp. 631–6. See also Chapter 2 of *The Poles*, by Willy Ley and the Editors of Life (Time-Life Books, 1966), for the details of innumerable expeditions which came to ruin while trying to reach the North Pole.
16. Jared Diamond, 'Paradises Lost', *Discover*, November 1997.
17. Ibid.
18. Ibid.
19. Peter F. Hamilton, *The Reality Dysfunction*, Macmillan, 1996. I owe these ideas also to discussions with Claes-Gustaf Nordquist, a keen student of projects for colonising space.
20. Michel Marriott, 'Ready to Replace Those Aging CD-Roms? DVD's Day is Dawning', *International Herald Tribune*, 7 May 1998.
21. This software can be obtained from Chris Naylor Research Ltd, 14 Castle Gardens, Scarborough, North Yorkshire, YO11 1QU, England.
22. Finney and Jones, op. cit., p. 155.
23. For a catalogue of existing expert systems, see Appendix C of Durkin's *Expert Systems: Design and Development*. For an account of how expert systems are programmed, the art of incorporating human expertise into software, see Feigenbaum and McCorduck's *The Fifth Generation*.

Chapter 2: STARSHIPS AND POLITICIANS

1. Edward Purcell, 'Radio Astronomy and Communication through Space', in Cameron (ed.), *Interstellar Communication*.

2. Bernal, *The World, the Flesh and the Devil*, p. 35.
3. Nicolson, *The Road to the Stars*, p. 201.
4. Lewis, *Mining the Sky*, Preface.
5. Peter M. Molton, 'On the Likelihood of a Human Interstellar Civilisation', *Journal of the British Interplanetary Society*, Vol. 31, 17 May 1978, pp. 203–8.
6. Obituary of Shizuo Takano, *Daily Telegraph*, 22 January 1992.
7. See Rachel Sylvester's report of a meeting between chiefs of the British Broadcasting Corporation and political party leaders, 'BBC Wants to Fade Out Boring Politicians', *Daily Telegraph*, 27 January 1998.
8. Frank Herbert, *Children of Dune*, quoted in Molton, op. cit.
9. Mallove and Matloff, *The Starflight Handbook*, pp. 257–8.
10. Molton, op. cit.
11. Ibid.

Chapter 3: THE TWILIGHT OF THE STATE

1. William Manchester, *A World Lit Only by Fire*, p. 9.
2. Ibid., p. 163.
3. As Paul Johnson relates in his *History of Christianity* (p. 269), by 1500 there were 73 presses in Italy, 51 in Germany, 39 in France, 24 in Spain, 15 in the Low Countries and eight in Switzerland. The most important of these, run by Aldus Manutius in Venice, was entirely devoted to publishing the recovered Greek classics.
4. Manchester, op. cit., p. 98.
5. Weatherford, *Savages and Civilisation*, p. 143.
6. I gave some account of how the RSA cipher works in my book *The Next 500 Years*. For a more authoritative explanation, see Martin E. Hellman, 'The Mathematics of Public-Key Cryptography', *Scientific American*, August 1979. See also Schneier's excellent *Applied Cryptography*.
7. Bruce Schneier, *Applied Cryptography*, pp. 124–6. Schneier calls this process 'unconditional sender and recipient untraceability'.
8. I owe this phrase, and the prediction behind it, to Davidson and Rees-Mogg, *The Sovereign Individual*, p. 243.
9. Simon Heffer, 'Why Shouldn't the Home Counties have Home Rule?', *Daily Mail*, 23 July 1997.
10. I have visited one such bank that offers this service, the network of the Mark Twain Banks, in St Louis, Missouri. See my article, 'The End of Money is Nigh: Encrypted Computer Cash Looks to be the Currency of the Future. And It Could Soon Be Impossible for Government to Tax Us', *Daily Telegraph*, 22 May 1996.
11. See two excellent articles on the social and economic consequences of trading on the Internet: Kelley Holland and Amy Cortese, 'The

Future of Money: E-cash Could Transform the World's Financial Life', *Business Week*, 12 June 1995; and 'The Disappearing Tax-payer', *The Economist*, 31 May 1997.

12. 'The Tax-Man's Nightmare', *Weekly Reuter Textline*, 24 February 1997.
13. Ian Johnson, 'In Need of Revenue, China Pushes Harder to Collect its Taxes', *Wall Street Journal Europe*, 11 June 1997.
14. Holland and Cortese, op. cit.
15. Davidson and Rees-Mogg, op. cit., pp. 83–4.
16. Andrew Alexander, 'I'm Not Really Laughing, Honest', *Daily Mail*, 6 June 1997.
17. Boris Johnson, 'France's Vote for the Euro', *Daily Telegraph*, 28 May 1997.
18. Moira Gunn, in *Wired*, July 1997.
19. John Laughland, in the *Daily Mail*, 28 May 1997.
20. All these facts are from Laughland's *Daily Mail* article cited above.
21. Joseph Fitchett, 'French Workaholics Beware: The Law Is Moving In', *International Herald Tribune*, 12 June 1988.
22. Susannah Herbert, 'Overdue and Over-Budget Shambles of New Library', *Daily Telegraph*, 21 July 1994.
23. Davidson and Rees-Mogg, op. cit., p. 101.
24. Ibid., pp. 101–2.
25. James Kim, 'Are Currency Traders Too Powerful?', *USA Today*, February 1998.
26. Gregory Millman, quoted by Kim, ibid.
27. Robert D. Kaplan, 'Was Democracy Just a Moment?' *Atlantic Monthly*, December 1997.
28. Ibid.
29. Millman, *The Vandals' Crown*, p. 63.
30. Grant N. Smith of the Millburn Ridgefield Corporation, quoted by Millman, op. cit., p. 64.
31. The estimate of $1.5 trillion worth of currency traded daily comes from *Time* magazine's cover story of Andrew Grove of Intel, 5 January 1998.
32. This estimate is based on figures for the gross world product compiled by the United Nations and the World Bank since 1972. For a discussion of these figures, which imply a doubling of human wealth every 25 years, see Chapter 2 of my book *The Next 500 Years*.
33. Quoted by Joe Flower, 'Off With Its Head: Because the Internet Symbolises Freedom of Information, Governments are Falling Over Themselves to Rein It In', *New Scientist*, 16 March 1996.
34. Geremie R. Barmé and Sang Ye, 'The Great Firewall of China', *Wired*, June 1997.
35. Ibid.

36. Flower, op. cit.
37. This is a most extreme – and hypothetical – example of the art of 'factorisation', breaking down numbers into their prime components. Most RSA users employ numerical keys of about 160 digits. In general, the time taken to factorise a number doubles each time a digit is added to it. In 1994, a team of mathematicians in 43 countries took nine months to discover that the number (known as RSA129, because it has 129 digits):

    114,381,625,757,888,
    867,669,235,779,976,
    146,612,010,218,296,
    721,242,362,562,561,
    842,935,706,935,245,
    733,897,830,597,123,
    563,958,705,058,989,
    075,147,599,290,026,
    879,543,541

    is divisible by the two primes:

    3,490,529,510,847,
    650,949,147,849,
    619,903,898,133,
    417,764,638,493,
    387,843,990,820,
    577

    and:

    32,769,132,993,266,
    709,549,961,988,
    190,834,461,413,
    177,642,967,992,
    942,539,798,288,
    533.

    In short, a key of 129 digits is much too small for a secure RSA cryptogram. The current minimum size in general use is 155 digits. The same methods and the same computers that were used to crack RSA129 would take about 50 million years to crack a 150-digit number provided that it was the product of two primes only and that they were both 'strong', i.e., sufficiently close to its square root.
38. P.W. Shor, in *Proceedings of the 35th Annual Symposium on the Foundations of Computer Science*, ed. S. Goldwasser (IEEE

Computer Society Press, Los Alamitos, California, 1994), p. 124. See also Tom Standage, 'The Coffee Cup Supercomputer: Quantum Computing', *Daily Telegraph* 'Connected', 3 June 1997. A full list of academic references to Shor's discovery can be found on the web site http://www.physik.uni-ulm.de/~sam/comp/node14.html.
39. Standage, op. cit. Good overviews of this coming technology are also given in C.H. Bennett, G. Brassard, and A.K. Ekert, 'Quantum Cryptography', *Scientific American*, October 1992; C. Zimmer, 'Perfect Gibberish', *Discover*, December 1992; and Ivars Peterson, 'Bits of Uncertainty; Blazing a Quantum Trail to Absolute Secrecy', *Science News*, 10 February 1996.
40. Adrian Berry, 'Cracking Good Code Trap Shuts Out Spies', *Sunday Telegraph*, 13 June 1993.

Chapter 4: THE MIGRATORY IMPERATIVE

1. The word is *jangala*, from which the English word jungle is derived. Gamble, *Timewalkers*, p. 197.
2. Numelin, *The Wandering Spirit*, p. 1.
3. Gamble, op. cit., pp. 8–9.
4. Jack London, *The Human Drift* (London, 1919), quoted by Gamble, op. cit., p. 15.
5. Rudyard Kipling, 'The Explorer'.
6. The German anthropologist P. Martin Gusinde, writing in 1926, quoted by Numelin, op. cit., p. 91.
7. Jeremiah, 35:7.
8. P.B. Du Chaillu, *Explorations and Adventures in Equatorial Africa*, London, 1861, p. 40.
9. Numelin, op. cit., p. 150.
10. William H. Prescott, *History of the Conquest of Mexico*, Vol. 1, London, 1847, pp. 15–16.
11. From Bradford's *History of the Plymouth Settlement*, pp. 21–2. Bradford was afterwards governor of Plymouth Colony in Massachusetts for thirty years.
12. J. Crawford, *A Dissertation on the Affinities of the Malayan Languages*, London, 1852.
13. John Lang, *View of the Origin and Migrations of the Polynesian Nation*, London, 1834.
14. Reader's Digest Association, *The Last Two Million Years*, p. 339. See also Jared Diamond, 'Paradises Lost', *Discover*, November 1997; also an excellent article on the Pacific Islands in *Encyclopedia Britannica*, Vol. 25, pp. 231–301.
15. Diamond, op. cit.
16. Ibid.
17. See, for example: P.S. Bellwood, 'The Peopling of the Pacific',

Scientific American, November 1980; William Howells, *The Pacific Islanders*, Weidenfeld and Nicolson, London, 1973; R.H. MacArthur and E.O. Wilson, *The Theory of Island Biogeography*, Princeton University Press, 1967.

18. Adrian Berry, 'If ET Really is Out There He Isn't Calling Home – Radio Messages from Aliens Fail to Pass the Intelligence Test', *Sunday Telegraph*, 26 September 1993. On the other hand, some astronomers fear that alien signals might not be reaching us because of their failure to penetrate interstellar gas clouds. See Marcus Chown, 'Brief Encounters: Clumpy Gas Between Stars Would Make any Signals from ET Fleetingly Short', *New Scientist*, 10 January 1998.
19. See, for example: Michael H. Hart, 'An Explanation for the Absence of Extraterrestrials on Earth', *Quarterly Journal of the Royal Astronomical Society*, Vol. 16, 1975, pp. 128–35; Frank J. Tipler, 'Extraterrestrial Intelligent Beings Do Not Exist', *Quarterly Journal of the Royal Astronomical Society*, Vol. 21, 1980, pp. 267–81.

    This is despite the UFO cults and such books as Erich von Daniken's *Chariots of the Gods* and its sequels, and Robert Temple's slightly more subtle (but not much) *The Sirius Mystery* (revised 1998), which claim that aliens have left their artefacts and/or knowledge on Earth. In the case of Temple and his belief that aliens arrived here from the Sirius system within the last 6,000 years, it is remarkable that he never asked radio astronomers to search Sirius for radio signals. Presumably the Sirius aliens, if they so recently had the technology to visit us, might still exist up there! See my hostile review of his 'improbable theorising', *Daily Telegraph*, 7 February 1998.
20. Andrew Lyne, quoted in Adrian Berry, 'A Billion Worlds for Man to Colonise', *Daily Telegraph*, 15 August 1991. Lyne was assuming the existence of about 100 billion stars in our Milky Way galaxy. Other estimates suggest a much greater number, of at least 250 billion, which would give us 2.5 billion Earth-sized planets in this galaxy. Of course, many of these planets would be uninhabitable because they would be at the wrong distances from their parent stars, or the stars themselves might be unsuitable. See Note 2 of Chapter 1.
21. Mallove and Matloff, *The Starflight Handbook*, p. 5.
22. Quoted by Mallove and Matloff, op. cit., pp. 1–2. I also quoted this hilarious passage in my book *The Next Ten Thousand Years*.
23. Konstantin Tsiolkovski (1857–1935), *Dreams of the Earth and the Sky*, 1895. The Soviet Government considered his work so important that they gave him a state funeral.
24. Dr Claes-Gustaf Nordquist, interview with the author.

25. An excellent account of the travails of the Huguenots, before and after the revocation of the Edict of Nantes, can be found in Sir Arthur Conan Doyle's historical novel *The Refugees*.

Chapter 5: THE STORY OF RIGIL KENT

1. *Brewer's Dictionary of Phrase and Fable* (1981), p. 214.
2. Savage, *The Millennial Project*, p. 332.
3. Within a region of 16 light years round the Sun (4,096 cubic light years) the German astronomer William Gliese has tracked 27 single stars, 14 double stars and five systems of triple stars, quoted in Couper and Henbest, *The Guide to the Galaxy*, p. 185.
4. Henry C. Ferguson, Nial R. Tanvir, and Ted von Hippel, 'Detection of Intergalactic Red-Giant-Branch Stars in the Virgo Cluster', *Nature*, Vol. 391, 29 January 1998, pp. 461–3.
5. A suggestion of Savage, op. cit., pp. 328–9.
6. I have taken the lore of Rigil Kent from R.H. Allen's *Star Names*, pp. 152–3, but even the learned Allen was unable to explain Toliman.
7. Father Richard, although by this achievement an important pioneer in astronomy, is almost unknown. I could discover nothing of his career, not even his first name. This anecdote appears on p. 172 of *The Photographic Atlas of the Stars*, by Arnold, Doherty, and Moore.
8. *Dictionary of National Biography*, Vol. 9, pp. 404–5.
9. Sir Fred Hoyle, *Astronomy and Cosmology: A Modern Course*, W.H. Freeman, San Francisco, 1975.
10. I have taken this simple experiment from James Pickering's *1001 Questions Answered About Astronomy*, Lutterworth Press, London, 1969, Answer No. 437, p. 137.
11. *Dictionary of Scientific Biography*, Vol. 7, pp. 17–18.
12. R. Innes, 'A Faint Star of Large Proper Motion', *Union Observatory Circular* (South Africa), No. 30, 1915.
13. Ibid.
14. Ken Croswell, 'Dealing with Problems on a First-Name Basis', *The Scientist*, Vol. 5, 11 November 1991, p. 13.
15. From the scientific Appendix to Pellegrino's novel *Flying to Valhalla*.
16. Ibid.
17. Sir John Herschel, *Outlines of Astronomy*, London, 1849.
18. Computer simulations of hypothetical planets in multistar systems by Robert and Betty Harrington, of the US Naval Observatory in Washington DC, and David Black, director of the Lunar and Planetary Institute in Houston, quoted by Pellegrino, op. cit.

19. 'Possibly First Planet Ever Seen Outside Solar System', Lee Gould, Associated Press, 28 May 1998.
20. Revd William Whewell, *A Dialogue on the Plurality of Worlds*, London, 1859.
21. Krauss, *Beyond Star Trek*, p. 54.
22. This would be the case with a hypothetical Earth-like planet orbiting the faint red star Ross 128, a star of an almost identical type to Lalande 21185. See Gillett's *World Building*, pp. 136–7; also Note 2 of Chapter 1 above.
23. Quoted in a 1997 University of Wisconsin web site on the search for alien planets, http://whyfiles.news.wisc.edu/017planet/main1.html.
24. Quoted in my article 'Jupiter, Saviour of the World', *Daily Telegraph*, 27 January 1996. Wetherill originally published his conclusions that giant gas planets were indispensable to a safe planetary system in 'Possible Consequences of Absence of "Jupiters" in Planetary Systems', *Astrophysics and Space Science*, Vol. 23, 1992, p. 212.
25. Here is a list of Jupiter-type planets found by the time of writing, August 1998. I give also their distances in light years to indicate the sensitivity of telescopes in the late nineties, and their power to observe the slight perturbations in the movements of far-off stars caused by the gravity of still unseen companions. This list is taken from Croswell, *Planet Quest*, pp. 210–11 (all their names are followed by the letter 'b' to distinguish them from their suns):

Planet	*Year found*	*No. of days in their year*	*Distance from star (AU)**	*Distance from Earth (light years)*
51 Pegasi b	1995	4.2	0.05	42
47 Ursae Majoris b	1995	1,100	2.1	42
Rho Cancri A b	1996	14.6	0.11	42
Tau Boötis A b	1996	3.3	0.05	52
Upsilon Andromedae b	1996	4.6	0.06	42
Lalande 21185 b	1996	13,000	9.5	8.3
Rho Coronae Borealis b	1997	39.6	0.23	52

* AU stands for 'astronomical unit', the distance between the Earth and the Sun, 150 million kilometres, a standard unit of distance in astronomy. Stating that 51 Pegasi b is 0.05 AU from its sun therefore means that this distance is only 7.5 million kilometres.

26. Gabrielle Walker, 'Giant Planet Hunted at Proxima Centauri', *Vancouver Sun*, 31 January 1998.
27. Savage, op. cit, p. 330.
28. Robert A.J. Matthews, 'The Close Approach of Stars in the Solar Neighbourhood', *Quarterly Journal of the Royal Astronomical Society*, Vol. 35, 1994, pp. 1–9.

Chapter 6: THE SIXTEEN SPEEDS

1. Arthur C. Clarke, in his *Profiles of the Future*, has a fascinating chapter called 'The Quest for Speed', outlining nine speed-bands (as opposed to my 16), in miles per hour, with the lowest between 1 and 10, and the highest between 100 million and 1,000 million.
2. J.D. Douglas (ed.), *The New Bible Dictionary*, Inter-Varsity Press, London, 1962, p. 370.
3. Isaiah, 31:1.
4. Deuteronomy, 17:16–17.
5. Joshua, 11:9; and II Samuel, 8:4. For an excellent discussion of the use of horses in the ancient world, see Juliet Clutton-Brock's *Horse Power: A History of the Horse and Donkey in Human Societies*, Natural History Museum Publications, London, 1962.
6. II Kings, 18:23; also Isaiah, 36:8.
7. Stephen Budiansky, *The Nature of Horses: Exploring Equine Evolution, Intelligence and Behaviour*, Free Press, New York, 1997. See an interesting review of this book in *Nature*, 27 March 1997.
8. Quoted in E.C. Krupp's fascinating article on cosmic horses, 'Horsefeathers on High', *Sky and Telescope*, January 1997.
9. Edward Gibbon, *The Decline and Fall of the Roman Empire*, Everyman's Library, Alfred A. Knopf, New York, Vol. 3, 1993, p. 11.
10. L. White, *Medieval Technology and Social Change*, Oxford University Press, 1962, quoted in Clutton-Brock, op. cit., p. 76.
11. William Manchester, *A World Lit Only by Fire*, p. 142.
12. W.H. Boulton, *The Pageant of Transport through the Ages*, p. 103.
13. Ibid., p. 121.
14. Ibid., p. 125.
15. One of a collection of amusing 'negative predictions' from the editorial 'Naysayers Never Die', *Aviation Week and Space Technology*, 22 March 1971. I used the anti-railway quotation of Van Buren (afterwards eighth President of the United States) in my 1974 book *The Next Ten Thousand Years*, but I take the liberty of using it again.
16. Robert Matthews, 'Mission Impossible', *Focus*, December 1996.
17. Marion C. Acworth, *The Great Delusion: A Study of Aircraft in*

Peace and War, written anonymously under the inexplicable pseudonym of 'Neon'.

18. Major Oliver Stewart, in the London *Morning Post*, 25 March 1926, quoted with approval by Acworth, op. cit.
19. This 1953 'speed graph' appears on p. 17 of Strong's *Flight to the Stars*. It was extrapolated from the speeds of sub-orbital rockets.
20. Quoted in Arthur Seldon and F.G. Pennance, *Everyman's Dictionary of Economics*, 1965.
21. Strong, op. cit., pp. 17–18.

Chapter 7: THE INCREDIBLE SHRINKING CALENDAR

1. The figures are reached by multiplying a given period on Earth by Einstein's formula: $(1-(v^2/c^2))^{1/2}$ where v is the speed of the ship in centimetres per second, and c is the speed in the same units. If, for example, the ship is travelling at 92 per cent of the speed of light, 60 minutes of Earth time will be compressed on board the ship by a factor of 0.392. Multiply this by 60 and we have 23.52 minutes.
2. Strong, *Flight to the Stars*, p. 139.
3. This story is told by G.J. Whitrow in his *What is Time?*, p. 7.
4. Quoted by J.B. Priestley, *Man and Time*, p. 166.
5. Newton's First Law states that an object continues on its course unless acted upon by a force, and his Third Law states that for every action – i.e., the backward thrust of a rocket engine, there is an 'equal and opposite reaction'.
6. George Gamow, *One . . . Two . . . Three . . . Infinity*, 1947.
7. Benedict Spinoza, *Ethics*, 1677. This saying is also attributed to Rabelais and to Cicero.
8. Plato, in his philosophical essay *Cratylus*.
9. Quoted in the *Encyclopedia Britannica*, 11th edition, 1911, p. 292.
10. Bernard Jaffe, *Michelson and the Speed of Light*, p. 81.
11. Analogy by Bertrand Russell, *The ABC of Relativity*, 1925.
12. Gardner, *Relativity for the Million*, p. 24.
13. Countless books have been written about how the Michelson–Morley experiment led to the special theory of relativity. By far the clearest and best, to my mind, is Gardner's *Relativity for the Million* (updated in 1997 as *Relativity Simply Explained*). See also an excellent CD-ROM, *A Brief History of Time: An Interactive Adventure*, based on Stephen Hawking's classic book with that title and published by Crunch Media, in association with W.H. Freeman, New York, and the *Scientific American*.
14. I owe these calculations to Savage, *The Millennial Project*, p. 322.
15. 'It is hard to know how they could have been accelerated to such terrific speeds or where they could have come from,' said physicist John Walker, writing in the summer issue of the *Newsletter of the*

Interstellar Propulsion Society. 'One possibility is that they were blasted out by the shock wave of a supernova explosion.'

16. These calculations are by Walker, op. cit. Walker has a fascinating Internet web site devoted to such topics as relativity theory and starship travel: http://www.fourmilab.ch. See also my article 'Super-Travel', *Astronomy Now*, October 1996.
17. Alexander Pope wrote the first couplet in the eighteenth century. Two hundred years later the journalist J.C. Squire answered him with the second.
18. From the hymn by Isaac Watts (1674–1748) that begins 'O God, our help in ages past . . .'

Chapter 8: THE ASCENT OF RIP VAN WINKLE

1. Strong, *Flight to the Stars*, p. 127. I quoted this passage with somewhat naïve approval in my *The Next 500 Years*.
2. Warren Salomon, 'The Economics of Interstellar Transport', *Analog*, May 1989. Reprinted in Schmidt and Zubrin, *Islands in the Sky*.
3. Radar observations of the asteroid 1986 DA reveal that it contains a virtual fortune in precious metals worth about $110 billion in mid-nineties prices. These include some 100,000 tons of platinum, worth $300 an ounce, a billion tons of nickel, valued at $10 a ton, and $100 billion worth of gold. Reported in *Science*, 3 June 1991. My book *The Next 500 Years* contains an extended discussion of the prospects for asteroid mining.
4. Salomon, op. cit.
5. Stefan Zweig, *Magellan: Pioneer of the Pacific*, London, 1938, p. 7.
6. The story of Rip van Winkle is told by Washington Irving in his *Sketchbook*, published in 1819.
7. Here is a simple BASIC computer program to calculate compound interest. The user has only to state the original amount of money, the rate of interest, and the number of 'periods' – i.e., years that it is invested:

```
10 INPUT"WHAT WAS THE ORIGINAL AMOUNT";A-
   MOUNT
20 INPUT"WHAT WAS THE RATE OF INTEREST";RATE
30 RATE=RATE/100
40 INPUT"WHAT WAS THE PERIOD";PERIOD
50 TOTAL=AMOUNT*(1+RATE)^PERIOD
60 PRINT"THE FINAL TOTAL IS";TOTAL
```

```
70 PAYOUT=TOTAL+AMOUNT
80 PRINT"WITH THE ORIGINAL INVESTMENT YOU
  RECEIVE";PAYOUT
```

Line 50 contains the key formula for calculating compound interest. The total interest payable equals the amount originally invested multiplied by the expression: 1 plus the interest rate, raised to the power of the number of periods. Please see the Glossary for an explanation of the difference between compound and simple interest.

8. I am grateful to financial experts Christopher Rawlings and Michael Tyrer for very kindly reading this chapter and making valuable comments. But I alone am responsible for any errors or misunderstandings.
9. Salomon, op. cit.

Chapter 9: A FUEL LIKE MAGIC

1. Edmond Rostand, *Cyrano de Bergerac*, trans. Brian Hooker, Bantam Books, New York, 1959 Act III, quoted by Mallove and Matloff, *The Starflight Handbook*, pp. 148–9.
2. I obtained these comparative efficiency figures for fission and fusion propulsion systems from Mallove and Matloff, op. cit., p. 44.
3. Savage, *The Millennial Project*, p. 316.
4. See the article on Paul Dirac in Isaac Asimov's *Biographical Encyclopedia of Science and Technology*, 1964.
5. Lapp, *Matter*, p. 154.
6. There is an important chapter on the nature and future uses of antimatter in Robert L. Forward's *Indistinguishable from Magic*.
7. Tim Radford, 'Pure destruction at heart of Milky Way', *Guardian*, 29 April 1997; Adrian Berry, 'Is this the engine for the 21st century?: The stupendous energy antimatter might yield', *Daily Telegraph*, 17 July 1989.
8. I owe these figures to Regis, *Nano*, p. 53.
9. Estimates by Robert L. Forward, quoted in my article, 'Is this the engine for the 21st century?', op. cit.
10. Savage, op. cit., p. 319.
11. Taking the predicted gross world product for the year 2000 as $17,000 billion. This estimate is based on figures produced by the United Nations and the World Bank which indicate that, for the past 20 years, and probably for much longer, human wealth has been increasing at an average annual rate of 2.8 per cent, thus doubling every 25 years. See 'The Wealth of the Race' in my book *The Next 500 Years*.

12. Forward, op. cit., p. 22.
13. Ibid., p. 26.
14. Ibid., p. 12.
15. Paul F. Massier, 'The Need for Expanded Exploration of Matter–Antimatter Annihilation for Propulsion Applications', *Journal of the British Interplanetary Society*, Vol. 35, September 1982, pp. 387–90.
16. George Chapline, 'Antimatter Breeders?', *Journal of the British Interplanetary Society*, Vol. 35, September 1982, pp. 423–4.
17. Massier, op. cit.
18. Robert L. Forward, quoted in my article, 'Is this the engine for the 21st century?', op. cit.
19. Lead has a melting point of 327.5° compared with daytime temperatures of 450° on Mercury. See an excellent article by Ken Croswell, 'Mercury – the Impossible Planet', *New Scientist*, 1 June 1991.
20. Isaac Asimov's short story 'Runaround', in *I, Robot*, Fawcett Publications, Greenwich, Connecticut, 1950.
21. From the ScienceNet web site: http://www.campus.bt.com/Campus . . .t/solarsys/mercuryteletext.html.
22. Croswell, op. cit.

Chapter 10: THE APPEARANCE OF A STARSHIP

1. Quoted by Sir Misha Black, 'The Aesthetics of Engineering', *Interdisciplinary Science Reviews*, Vol. 1, No. 1, 1976.
2. A study by Donald Goldsmith and Tobias Owen, quoted in the scientific Appendix of Pellegrino's *Flying to Valhalla*.
3. This picture can be seen on p. 63 of Arthur C. Clarke's *Man and Space*.
4. Clarke, op. cit., p. 70.
5. I owe this idea, and much that follows, to Pellegrino's Appendix to *Flying to Valhalla*.
6. Kaku, *Visions*, p. 304.
7. In addition to the K. Eric Drexler classic *Engines of Creation*, and Ed Regis's excellent book *Nano*, two recent popular articles describe the principles of nanotechnology: Emma Bayley, 'Nanotechnology: The Diamond Age', *Focus*, April 1998, pp. 21–2; and Will Hively, 'The Incredible Shrinking Finger Factory', *Discover*, March 1998, pp. 84–93.
8. Pellegrino, op. cit.
9. The Chemical Rubber Company, *Handbook of Chemistry and Physics* (1981), p. B-44; *Encyclopedia Britannica* (1994), Vol. 21, pp. 461 and 464–5. Tungsten is known in some countries as wolfram, and its chemical symbol is W.

10. *Encyclopedia Britannica*, Vol. 12, p. 41. See also Grant Heiken, David Vaniman and Bevan M. French, *Lunar Sourcebook: A User's Guide to the Moon*, Cambridge University Press, 1991, p. 413.
11. Pellegrino, op. cit.
12. The density of air molecules at sea level on Earth is about 10^{-3} grams per cubic centimetre compared with around 10^{-23} grams per cubic centimetre for interstellar space in our region of the galaxy. See, for example, C.W. Allen, *Astrophysical Quantities*, p. 263, who stresses that our current estimates of the latter are still approximate and uncertain.
13. Pellegrino, op. cit.
14. Ibid.
15. Mallove and Matloff, *The Starflight Handbook*, p. 161.
16. Ibid., p. 170.
17. Some atoms, such as those of potassium, sodium, and caesium, possess such low 'ionisation energies' that they can be turned into plasma at temperatures as low as 3,000 degrees. But others, particularly the carbon, silicon, iron, and magnesium that are the most important components of interstellar dust grains, will become plasma at temperatures of about 10,000 degrees. *Encyclopedia Britannica*, Vol. 23, p. 666.
18. Pellegrino, op. cit. He adds: 'When a dust grain impacts against the droplets, the particles from which the grain is made will simply act as individual particles, "unaware" that they are part of anything else. Hence, as it penetrates the droplet, each proton, neutron or electron will be scattered at a certain angle, and the angle of scatter should increase as the particles pass through more and more droplets in their approach to the ship's magnetic field, resulting in a harmless, spreading shower effect.'
19. James Strong believes that eight to ten years would be necessary for flight-testing, a period three times longer than the actual voyage to Alpha Centauri! Strong, *Flight to the Stars*, p. 41.
20. Pellegrino, op. cit.
21. I owe the idea of using a sail to brake the starship to my good friend and scientific adviser Claes-Gustaf Nordquist (private correspondence).
22. In 1994 the Europe–NASA *Ulysses* spacecraft measured these differing speeds of the solar wind. See Adrian Berry, 'Spacecraft Reports on Sun's Winds of Change', *Daily Telegraph*, 16 September 1994.
23. Zubrin, *The Case for Mars*, p. 243.
24. As described in the twelfth-century *Saga of King Olaf*.
25. Zubrin, op. cit., p. 244.
26. Ibid, pp. 243–5. See also R. Zubrin and D. Andrews, 'Magnetic Sails and Interplanetary Travel', *Journal of Spacecraft and Rockets*,

April 1991. For a general overview of the science of magnetic fields and electric currents, see *Encyclopedia Britannica*, Vol. 18, pp. 175–9.

27. Zubrin, op. cit., pp. 244–5.
28. Mauldin, on p. 89 of his *Prospects for Interstellar Travel*, states flatly that the sail and similar braking systems are 'far too weak' to slow the ship from relativistic speeds. But other experts whom I have consulted are just as insistent that they will work.
29. *Architects' Journal*, 6 March 1974, quoted by Black, op. cit.

Chapter 11: ROCKETLESS ROCKETRY – AND OTHER METHODS

1. Mauldin, *Prospects for Interstellar Travel*, p. 59.
2. 'He had been eight years on a project for extracting sun-beams from cucumbers, which were to be put into vials hermetically sealed, and let out to warm the air in raw inclement summers.' Jonathan Swift, *Gulliver's Travels*, Ch. 5.
3. Mauldin, op. cit., p. 87. Also see Gregory Matloff, 'Faster Non-Nuclear World Ships', *Journal of the British Interplanetary Society*, Vol. 39, 1986, pp. 475–85.
4. A suggestion by Mauldin, op. cit., p. 89.
5. Ibid., p. 88.
6. For a good round-up of these alternative ideas, see Robert L. Forward, 'Ad Astra!', *Journal of the British Interplanetary Society*, Vol. 49, 1996, pp. 23–32. Also R.M. Zubrin and D.G. Andrews, 'Magnetic Sails and Interplanetary Travel', *Journal of Spacecraft and Rockets*, Vol. 28, 1991, pp. 197–203.
7. Ken Croswell, 'Interstellar Trekking', *Astronomy*, June 1998.
8. Alan Bond and Anthony Martin, 'Project Daedalus: Introduction and Mission Profile', special supplement to the *Journal of the British Interplanetary Society*, 1978, pp. S5–S8 and pp. S37–S42.
9. R.W. Bussard, 'Galactic Matter and Interstellar Flight', *Astronomica Acta*, Vol. 6, 1960, pp. 179–94.
10. A.R. Martin, 'Some Limitations of the Interstellar Ramjet', *Spaceflight*, Vol. 14, February 1973, pp. 21–5.
11. Alan Bond, 'Analysis of the Potential Performance of a Ram-Augmented Interstellar Rocket', *Journal of the British Interplanetary Society*, Vol. 27, 1974, pp. 674–85.
12. For discussion of a boron–lithium reaction in a starship engine, see Mallove and Matloff, *The Starflight Handbook*, pp. 116–19.

Chapter 12: THE STAR ON THE STARBOARD BEAM

1. See Gary Grittner, 'Tell Time by the Big Dipper: Who Needs a Watch when the Sky Tells Us What Time it is on Every Clear Night?', *Astronomy*, April 1997.
2. This scheme of navigation is described by James R. Wertz, 'Interstellar Navigation', *Spaceflight*, Vol. 14, June 1972, pp. 206–16.
3. This is no exaggeration. Using the formula $\frac{4}{3}\pi r^3$ for the volume of a sphere, I divided Earth's volume of 10^{12} cubic kilometres into that of Betelgeuse, 3.8×10^{25} cubic kilometres, getting an answer of 38×10^{12}, or 38 trillion.
4. There is an excellent description of the Doppler–Fizeau effect in Asimov's *A Choice of Catastrophes*.
5. The aft view is described thus by two experts: 'At these extreme speeds only the most extreme ultra-violet sources, with appreciable emission intensity in the X-ray, and/or gamma ray region of the electromagnetic spectrum, could offer a visible aft-image after undergoing extreme red-shift of wavelength.' R.W. Stimets and E. Sheldon, 'The Celestial View from a Relativistic Starship', *Journal of the British Interplanetary Society*, Vol. 34, 1981, pp. 83–99.
6. Nicolson, *The Road to the Stars*, pp. 116–17.
7. About 41 seconds of arc during the course of a year. I owe this concise description to Patrick Moore's *Atlas of the Universe*. For other good discussions of stellar aberration, see Mallove and Matloff, *The Starflight Handbook*, and Nicolson, op. cit.
8. The maximum displacement of stars by stellar aberration is calculated – in radians – by the formula $2v/c$, where v is the speed of the observer and c is the speed of light. One must then multiply the answer by 57.3 to convert radians into angles. As seen from Earth, the displacement is very small: $2v/c \times 53.3$ yields 41 arc seconds, or about 0.011 of an angle, since an arc second is $1/3{,}600$ of an angle.
9. Please see Appendix III for the computer program 'Navigating a Starship at Close to the Speed of Light'.
10. Stimets and Sheldon, op. cit.
11. Ibid., quoted by Mallove and Matloff, op. cit., Ch. 12.
12. Mallove and Matloff, op. cit., p. 185.

Chapter 13: FASTER THAN LIGHT?

1. Isaac Asimov, *The Stars Like Dust*, Panther Books, London, 1958.
2. Kaku, *Hyperspace*, p. 244.
3. A. Einstein and N. Rosen, 'The Particle Problem in the General Theory of Relativity', *Physical Review*, Vol. 48, 1 July 1935, pp. 73–7. See also A. Einstein and E.G. Strauss, 'The Influence of the Expansion of Space on the Gravitational Fields Surrounding the

Individual Stars', *Reviews of Modern Physics*, Vol. 17, nos. 2 and 3, April–July 1945, pp. 120–4; Rees, Ruffini and Wheeler, on p. 217 of their *Black Holes, Gravitational Waves and Cosmology*, characterise this latter paper as describing the 'Swiss cheese universe'.

4. John A. Wheeler and Seymour Tilson, 'The Dynamics of Space-Time', *International Science and Technology*, December 1963.
5. M.S. Morris and K.S. Thorne, 'Wormholes in Spacetime and their Use for Interstellar Travel: A Tool for Teaching General Relativity', *American Journal of Physics*, Vol. 56, 1988, p. 411.
6. Quoted by Jeff Greenwald, 'To Infinity and Beyond', *Wired*, July 1998. Thorne added: 'We physicists have tended to avoid such questions because they are so close to science fiction. While many of us may enjoy reading science fiction or even write some, we fear ridicule from our colleagues for working on research close to the science fiction fringe.'
7. Adrian Berry, 'Cosmic Wormholes: Key to the Universe: A Blueprint for Instantaneous Travel Anywhere in Space', *Daily Telegraph*, 5 December 1989.
8. *Encyclopedia Britannica*, Vol. 7, pp. 826–7.
9. F. Tipler, 'Causality Violation in Asymptotically Flat Space-Times', *Physical Review Letters*, Vol. 37, 1976, p. 979, quoted by Kaku, op. cit., p. 244.
10. Quoted by Heinz Pagels, *The Cosmic Code*, Bantam Books, New York, 1982, pp. 173–4. See also Kaku, op. cit., p. 315.
11. As I say, my explanation is highly simplified, probably over-simplified. More complete versions can be found in: Alan Guth, 'Cooking up a Cosmos: Will Our Descendants Create a Universe in the Laboratory?', *Astronomy*, September 1997; Andrei Linde, 'The Self-Reproducing Inflationary Universe', *Scientific American Quarterly* ('The Magnificent Cosmos'), Spring 1998; John G. Cramer, 'Other Universes I', *Analog Science Fiction & Fact Magazine*, September 1984; and Fraser, Lillestøl and Sellevåg, *The Search for Infinity*, pp. 102–5.
12. For an interesting discussion, see Coveney and Highfield, *The Arrow of Time*.
13. Barrow, *Impossibility*, p. 200.
14. Kaku, op. cit., p. 235.
15. Quoted by Kaku, op. cit.
16. David Deutsch, quoted by Barrow, op. cit., p. 207. See also Deutsch, *The Fabric of Reality*.
17. For an account of this meeting, see John G. Cramer, 'NASA Goes FTL Part 1: Wormhole Physics', *Analog Science Fiction & Fact Magazine*, Mid-December 1994; and 'NASA Goes FTL Part 2: Cracks in Nature's Armour', *Analog Science Fiction and Fact*

Magazine, February 1995. Cramer is a professor of physics at the University of Washington in Seattle, and a distinguished science fiction writer.
18. Tachyons were first proposed by Gerald Feinberg in his article 'Particles That Go Faster Than Light', *Scientific American*, February 1970. Nobody has found one at the time of writing.
19. Arthur C. Clarke, 'A Walk in the Dark', 1945.
20. M. Alcubierre, 'The Warp Drive: Hyper-Fast Travel Within General Relativity', *Classical and Quantum Gravity*, Vol. 11, 1994, pp. L73–L77.
21. John G. Cramer, 'The Alcubierre Warp Drive', *Analog Science Fiction and Fact Magazine*, November 1996.
22. See, for example, John Horgan, *The End of Science: Facing the Limits of Knowledge in the Twilight of the Scientific Age*, Helix Books, New York, 1997. I did not find Horgan's argument at all convincing.

Chapter 14: 'I ENJOY WORKING WITH PEOPLE'

1. F.J. Tipler, 'Extraterrestrial Beings Do Not Exist', *Quarterly Journal of the Royal Astronomical Society*, Vol. 21, 1980, pp. 267–81. From its title, this paper might not appear to have much to do with computers. But its thesis is that with self-reproducing machines – computers that could make identical copies of themselves – aliens could have colonised the galaxy by jumping from star to star. They have not done so; therefore, in Tipler's view, they do not exist.
2. A calculation by Strong, *Flight to the Stars*, pp. 103–4.
3. This remarkable phrase I owe to Ravishankar K. Iyer, of the University of Illinois at Urbana, Champaign. But I have not been able to discover which Byzantine generals had this problem. See Iyer's brilliant essay: ' "Foolproof and Incapable of Error?" Reliable Computing and Fault Tolerance', a contribution to *HAL's Legacy*, ed. David G. Stork.
4. How does Poole get into this appalling position? His play goes wildly wrong after the 11th move (see the game listing below), when HAL successfully lays a trap for him. When HAL takes one of Poole's Knights in the 11th move, exposing a Rook on the eighth square, Poole cannot resist taking it, even though the Rook is no threat to him, thereby immobilising his Queen! HAL now advances his Queen and a Bishop, and Poole is finished. For a much more profound analysis of this game than I am capable of giving, see ' "An Enjoyable Game": How HAL Plays Chess', a contribution by Murray S. Campbell, a chess computer scientist at IBM, to Stork, op. cit.

White (Poole)	*Black (HAL)*
1. P-K4	P-K4
2. Kt-KB3	Kt-QB3
3. B-B5	RP-P3
4. B-R4	Kt-KB3
5. Q-K2	P-QKt4
6. B-Kt3	B-K2
7. P-B3	Castles
8. Castles	QP-P4
9. PxP	KtxP
10. KtxP	Kt-B6
11. Q-Q4	KtxKt
12. QxR	Q-Q6
13. B-Q1	B-R6
The film dialogue starts here:	
14. QxP	BxP
15. BxQ	Kt-R6 mate

5. Ibid.
6. Ibid.
7. Ibid.
8. See my article on Lenat's efforts to create HAL: 'Mind-blowing computer is a reasonable prospect: They pulled his plug in *2001*, but we haven't heard the last of HAL', *Sunday Telegraph*, 5 November 1995.
9. Douglas Lenat, 'From 2001 to *2001*: Common Sense and the Mind of HAL', contribution to Stork, op. cit.
10. From Ed Regis, *Who Got Einstein's Office? Eccentricity and Genius at the Institute for Advanced Study*, Addison-Wesley, Reading, Massachusetts, 1987.
11. See a description of the once mighty ILLIAC 4 in David Kuck's 'Can We Build HAL?', a contribution to Stork, op. cit.
12. To be exact, Moore's Law states that computing speeds and the memory-density of chips doubles every 18 months. For simplicity, I have rephrased this as 'processing power'.
13. For a good account of progress in teaching machines to speak, see Joseph Olive's 'The Talking Computer: Text to Speech Synthesis', his contribution to Stork, op. cit.
14. Raymond Kurzweil, 'When Will HAL Understand What We Are Saying? Computer Speech Recognition and Understanding', contribution to Stork, op. cit.
15. Lenat, op. cit.

16. Roger Penrose, *The Emperor's New Mind: Concerning Computers, Minds, and the Laws of Physics*, 1989.
17. A.M. Turing, 'Computing Machinery and Intelligence', *Mind: A Quarterly Review of Psychology and Philosophy*, Vol. 59, October 1950, p. 236.
18. I very much owe this idea to Kevin Warwick, an expert in machine intelligence at Reading University.
19. Quoted by John Markoff, 'A Subtle Artisan Finds a Medium in New Analogue Chips', *New York Times*, 27 July 1998.
20. See the chapter entitled 'Heavenly Computers' in my book *Galileo and the Dolphins*. McLaughlin wrote his original suggestion in the journal *Spaceflight*.
21. Ambrose Bierce, 'Moxon's Master', from *The Collected Works of Ambrose Bierce*, Citadel Press, New York, 1970.

Chapter 15: WHEN LUXURY IS NECESSITY

1. From a 1994 study by Harvard University School of Public Health. Quoted by Diane Fairechild, a flight attendant for 21 years and author of *Jet Smart* (a book about healthy flying) on her web site: http://www.flyana.com/full.html. See also Sharon Davey, 'False Economy', *Sunday Times*, 22 June 1997; Michael Conlon, 'Exercises at 30,000 Feet', *Reuters*, 2 October 1997.
2. Laurence Moyer, *Victory Must be Ours: Germany in the Great War 1914-1918*, Hippocrene Books, New York, 1995. The passage is based on reports in the *New York Times* of 16 and 22 May 1914.
3. For a fascinating discussion of artificial gravity in a spacecraft, see 'Artificial Gravity: Which Way is Up?' by John G. Cramer of *Analog Science Fiction and Fact Magazine*, on his web site at www.npl.washington.edu/AV/altvw18.html. Cramer calculates that to create Earth-normal gravity in the outer regions of a spacecraft, a vessel with a diameter of 160 metres should make a full rotation every 18 seconds.
4. I have taken this distinction from Asimov, *The World of Carbon*, p. 9.
5. Ibid., p. 12.
6. These examples are taken from Sally and Lucian Berg's *New Foods for All Palates: A Vegetarian Cook Book*, Gollancz, London, 1973, p. 179.
7. From Richard Feynman's seminal lecture, 'There's Plenty of Room at the Bottom', at the 1959 annual meeting of the American Physical Society. Reprinted in *Engineering and Science*, Vol. 23, February 1960, p. 22; and in the book *Miniaturisation*, ed. H.D. Gilbert, Reinhold, New York, 1961; summarised in Regis, *Nano*, Ch. 4.

8. Regis, op. cit., Prologue 1. See also John Walker, 'The Coming Revolution in Manufacturing', on his web site www.fourmilab.ch.
9. Regis, op. cit., p. 23.
10. Drexler, *Engines of Creation.*
11. Regis, op. cit., p. 6.
12. Quoted by Susan Maclean Kybett, *History Today*, October 1989.
13. I owe the word 'corpsicles' to Lewis, *Mining the Sky*, p. 245.
14. I owe these ideas to a fellow interstellar travel enthusiast, Dr Claes-Gustaf Nordquist, former Chief Medical Officer of the Swedish Royal Horse Guards and the Royal Household Brigade of the Royal Swedish Army.
15. Rachel Kelly, 'Let There be Light for Winter', *The Times*, 16 November 1996.
16. See the chapter entitled 'Holodecks and Holograms' in Krauss, *The Physics of Star Trek.*

Chapter 16: THE GAME'S AFOOT, WATSON

1. Robert Naeye, 'Winter at the Pole', *Astronomy*, May 1997. Interview with John Briggs of Yerkes Observatory, who wintered at the South Pole in 1994.
2. Rachel Preiser, 'Personal Space', contribution to the article 'Castle in the Air', *Discover*, May 1997.
3. Ibid.
4. Ibid.
5. This description of Heinlein's novel is given by John Clute and Peter Nicholls on p. 480 of their *Encyclopedia of Science Fiction.*
6. Adrian Berry, 'To Mars by way of the Caribbean', *Daily Telegraph Magazine*, 17 July 1970.
7. A reminiscence of Marc Saltzman, 'In the Game', *Toronto Star*, 27 February 1997.
8. *Sherlock Holmes and the Case of the Rose Tattoo* has nothing to do with any written Holmes adventure written by Conan Doyle. It is one of a series called *The Lost Files of Sherlock Holmes*, and was released by Electronic Arts, of Slough, England, in 1996.
9. See, for example, Adrian and Marina Berry, 'Solving a Dastardly Crime', *The Spectator*, 3 May 1997.
10. Jane Jensen, creator of *Gabriel Knight 2: The Beast Within.* See following Note.
11. *Gabriel Knight 2: The Beast Within* was released by Sierra On-Line in 1995.
12. Some of the keystrokes needed to create this amateurish drawing are not actually on a PC's keyboard, but are available when using a good word processor. With WordPerfect version 6 I created the necessary ASCII characters by holding down the ALT key while

typing numbers on the right numberpad. I created the 'gun' with ASCII character 204, and the ringed R and C with the respective ASCII characters 169 and 184. ASCII, short for American Standard Code for Information Interchange, is a code in which all the conceivable characters that a computer keyboard can produce (255 of them) is each assigned a number.

13. This advance is enabling 'multi-disk' games, such as *The Beast Within*, to be sold more conveniently for the player on a single disk.
14. For a demonstration of this technology, see the program *Video Reality*, distributed by SouthPeak Interactive LLC, of Cary, North Carolina.
15. For a description of Plots Unlimited, and how the writer Arthur Weingarten has used it to help him construct Inspector Maigret detective stories, see the chapter entitled, 'The Day the Computer Murdered the Magnate' in my book *Galileo and the Dolphins*.
16. For more information about The Collaborator, see 'The Day the Computer Murdered the Magnate', op. cit.
17. A prediction by Paul Keery, 'Maybe We Don't Need the Players at all Any More', *Toronto Star*, 6 October 1994.
18. For a good discussion of radioneurology and radiotelepathy, see Freeman Dyson, *Imagined Worlds*, pp. 95–137.
19. Ibid., p. 135.

Chapter 17: THE BIG SLEEP

1. This list is based on Dole, *Habitable Planets for Man*, published in 1970. A longer and more recent list, showing 28 stars out to a distance of 21 light years that might have habitable planets, is given by M.J. Fogg, 'An Estimate of the Prevalence of Biocompatible and Habitable Planets', *Journal of the British Interplanetary Society*, January 1992.
2. Berry, *Galileo and the Dolphins*, pp. 203–4.
3. Lapp, *Matter*, p. 62.
4. *Guinness Book of Records* (1995), p. 22.
5. Lapp, op. cit., p. 62.
6. Du Charme, on p. 144 of his book *Becoming Immortal*, lists these organisations, with their addresses and telephone numbers, as: the Alcor Life Extension Foundation in Scottsdale, Arizona; the American Cryonics Society in Cupertino, California; Cryocare in Culver City, California; the Cryonics Institute in Detroit, Michigan; the International Cryonics Foundation in Stockton, California; and Trans Time Inc., in Oakland, California. Interestingly, Alcor recently moved its storage facilities from California to Arizona for fear of future earthquakes. See a discussion of the aims of Cryocare in Steven B. Harris, 'Many are Cold but Few are Frozen: A

Physician Considers Cryonics', 1995, published on the Internet at http://www.cryocare.org.cryocare/humanist/html.
7. Gerry Byrne, 'The Eternal Problem', *Electronic Telegraph*, 22 January 1998.
8. Ralph C. Merkle, 'Cryonics, Cryptography, and Maximum Likelihood Estimation', a paper published in *Proceedings of the First Extropy Institute Conference*, a meeting held in Sunnyvale, California, in 1994. It can be found on the Internet at http://merkle.com/merkle.cryptoCyro.html.
9. Ibid.
10. For an overview of this still non-existent technology, see Ralph C. Merkle, 'The Molecular Repair of the Brain', an article published in two parts in *Cryonics* magazine, Vol. 15, Nos. 1 and 2, January and April 1994. A shorter version of this article, entitled 'The Technical Feasibility of Cryonics', appeared in *Medical Hypotheses*, Vol. 39, 1992, pp. 6–16.
11. Du Charme, op. cit., pp. 146–7.
12. Ibid., p. 147.
13. See, for example, Regis, *Nano*.
14. DNA from the magnolia leaves was between 17 and 20 million years old. See Edward M. Golenberg and others, 'Chloroplast DNA Sequence from a Miocene Magnolia Species', *Nature*, Vol. 34, 12 April 1990, p. 656, quoted by Merkle, 'The Molecular Repair of the Brain', op. cit.
15. A list of the 563 stars with actual names, most of them Arabic, can be found in Michael Bakich, *The Cambridge Guide to the Constellations*, pp. 111–25. The vast majority of unnamed stars have the prefix HD followed by a number designating them in the *Henry Draper Catalogue* of 1,072,000 stars, last updated in 1930.

Chapter 18: THE FADING OF THE STARS

1. William Shakespeare, *Julius Caesar*, Act II, Scene 2.
2. For an excellent general description of comets, see Moore, *Atlas of the Universe*, pp. 126–31.
3. Scott's painting of the 1759 appearance of Halley's Comet is reproduced in Rowan-Robinson, *Our Universe*. Reproduced again on the science page of the *Daily Telegraph* on 1 October 1990, the original is in the hands of private owners who, for security reasons, cannot be identified.
4. Quoted by Moore, op. cit., p. 128.
5. See the obituary of Jan Oort (1900–92), *Daily Telegraph*, 13 November 1992.
6. Freeman Dyson, *Disturbing the Universe*, p. 236.

7. Dorling Kindersley, CD-ROM *Encyclopedia of Space and the Universe.*
8. Dyson, op. cit., p. 236.
9. Moore, op. cit., p. 128.
10. Assuming, for convenience, that the comet's nucleus is an exact sphere with a radius of 25 kilometres, its surface area will be $4\pi r^2$, or 7,854 square kilometres.
11. See the chapter entitled 'Greening the Galaxy', in Dyson, op. cit. Also Chapters 6 and 7 of Savage, *The Millennial Project*; and Ben R. Finney and Eric M. Jones, 'Fastships and Nomads', a contribution to their jointly edited book *Interstellar Migration and the Human Experience.*
12. David G. Stephenson, 'Comets and Interstellar Travel', *Journal of the British Interplanetary Society*, Vol. 36, 1983, pp. 210–14, quoted by Savage, op. cit., p. 305.
13. Dyson, 'Greening the Galaxy', op. cit.
14. For discussion on this point, see F.J. Dyson, 'The Search for Extraterrestrial Technology', a contribution to Marshak, *Perspectives in Modern Physics.*

Chapter 19: A COSMIC INTERNET

1. For a description of this signal, see Couper and Henbest, *To the Ends of the Universe*, pp. 112–113.
2. Ben Iannotta, 'Earth, You've Got Mail', *New Scientist*, 22 May 1999.
3. Ibid.
4. Ferris, *Coming of Age in the Milky Way*, p. 376.
5. Jack Littlefield, 'Melancholy Notes on a Cablegram Code Book', *The New Yorker*, 28 July 1934. Quoted by Kahn, *The Codebreakers*, pp. 851–3.
6. Ferris, op. cit., p. 379.
7. Ferris, op. cit.
8. Ferris, op. cit.

Chapter 20: EXTINCTION AND ETERNITY

1. Jeffrey Winters, 'A Brief Tour of a Bad Cosmic Neighbourhood', *Discover*, April 1998. See also Chapter 6 of Couper and Henbest, *The Guide to the Galaxy.*
2. Erik M. Leitch and Gautam Vasisht, 'Mass Extinctions and the Sun's Encounters with Spiral Arms', *New Astronomy* (an on-line journal, at www.elsevier.nl/locate/newast), Vol. 3, 1998, pp. 51–6.
3. Jeff Hecht, 'Fiery Future for Planet Earth', *New Scientist*, 2 April 1994; Malcolm W. Browne, 'Scientists say Sun will Melt Earth; in 1.1 Billion Years, Expanding Star will Turn Our Planet into Space Cinder', *Ottawa Citizen*, 20 September 1994.

4. Moore, *Atlas of the Universe*, p. 140.
5. Strange to say, *The Time Machine* is *not* to be found in any edition of Wells's complete works. For some reason he took a dislike to the novel, and set out to destroy every copy he could find. Thankfully he didn't do a very good job. This quotation comes from *The Wheels of Chance and The Time Machine*, Everyman's Library, 1935.
6. Hecht, op. cit.; Browne, op. cit.
7. Lorne Whitehead, 'Light Elements', *Discover*, July 1996. Also, Adrian Berry, 'Future Vision of a Man Who Would Move the Earth', *Electronic Telegraph*, 15 July 1996.
8. Whitehead, op. cit.
9. Quoted by Browne, op. cit.
10. Fred C. Adams and Gregory Laughlin, 'A Dying Universe: The Long Term Fate and Evolution of Astrophysical Objects', *Reviews of Modern Physics*, April 1997. Adams and Laughlin also presented their paper at the 1997 winter meeting in Toronto of the American Astronomical Society. See also Adams and Laughlin, 'The Future of the Universe', *Sky and Telescope*, August 1998.
11. Joseph Hall, 'The End of the World – but Don't Worry, It's a Very, Very Long Way Off', *Toronto Star*, 16 January 1997.
12. Robert Matthews, 'To Infinity and Beyond', *New Scientist*, 11 April 1998. Also, Kathy Sawyer, 'Universe Will Keep Expanding For Ever, Research Teams Say', *Washington Post*, 9 January 1998.
13. Adams and Laughlin, 'A Dying Universe', op. cit. They also mention the alternative possibility that the Earth may escape this fate, being pushed into a much wider orbit by the immeasurably stronger solar winds of the dying Sun.
14. *Oxford Dictionary*, Vol. 3, p. 10.
15. Adams and Laughlin, 'A Dying Universe', op. cit.
16. It is at this point that Adams and Laughlin differ from an earlier, much more optimistic study by Freeman Dyson, who believed that proton decay would not take place, and that there would consequently be a far longer period in which life would be possible. See F.J. Dyson, 'Time Without End: Physics and Biology in an Open Universe', *Reviews of Modern Physics*, Vol. 51, No. 3, July 1979, pp. 13–18. I discussed Dyson's ideas in the Introduction to my *The Next 500 Years*. It is hard for a layman to tell who is right in this matter, but proton decay does seem to be predicted by the Second Law of Thermodynamics.
17. Fraser, Lillestøl and Sellevåg, *The Search for Infinity*, p. 84.
18. Ibid., pp. 84–5.
19. This idea is explained in Chapter 7, 'Black Holes Ain't So Black', of Stephen Hawking, *A Brief History of Time*. See also Fraser, Lillestøl and Sellevåg, op. cit., pp. 118–19.

20. Paul Recer, 'Astronomers say Universe will Become Abysmally Black, Lifeless', *Associated Press*, 15 January 1997.
21. A summary which I have based fairly loosely on that by Adams and Laughlin, 'A Dying Universe', op. cit., omitting many stages.
22. Since the period from the present epoch to the end of the Stelliferous Era is 10^{14} years, the sum is $10^{14} \times 100 \div (10^{99} \times 10)$, producing an answer of 10^{-84}, the extra '× 10' because of the limitations of pocket calculators.

Bibliography

In the case of two or more authors or editors, the book is listed under whichever is first in alphabetical order.

ADAMS, Fred. With Greg Laughlin. *The Five Ages of the Universe: Inside the Physics of Eternity* (The Free Press, New York, 1999).

ALEXANDER, William. With Arthur Street. *Metals in the Service of Man* (Penguin Books, London, 1989).

ALLEN, C.W. *Astrophysical Quantities* (Athlone Press, London, 1973).

ALLEN, Richard Hinckley. *Star Names: Their Lore and Meaning* (Dover Publications, New York, 1994; originally published 1897).

ARNOLD, H.J.P. With P. Doherty and P. Moore. *The Photographic Atlas of the Stars* (Institute of Physics, Bristol, 1997).

ASHPOLE, Edward. *Where is Everybody? The Search for Extraterrestrial Intelligence* (Sigma Press, Wilmslow, Cheshire, 1997).

ASIMOV, Isaac. *A Choice of Catastrophes* (Hutchinson, London, 1980).

—*The World of Carbon* (Abelard-Schuman, New York, 1958).

—*Asimov's Biographical Encyclopedia of Science and Technology: The Living Stories of more than 1,000 Great Scientists from the Age of Greece to the Space Age, Chronologically Arranged* (Doubleday, New York, 1964).

BAKER, H. Robin. *Migration: Paths Through Time and Space* (Hodder and Stoughton, London, 1982).
BAKICH, Michael E. *The Cambridge Guide to the Constellations* (Cambridge University Press, 1995).
BALL, Philip. *Made to Measure: New Materials for the 21st Century* (Princeton University Press, 1997).
BARROW, John D. *Impossibility: The Limits of Science and the Science of Limits* (Oxford University Press, 1998).
—With Frank J. Tipler. *The Anthropic Cosmological Principle* (Oxford University Press, 1996).
BERNAL, J.D. *The World, the Flesh and the Devil: An Enquiry into the Three Enemies of the Rational Soul* (London and New York, 1929).
BERRY, Adrian. *The Next Ten Thousand Years: A Vision of Man's Future in the Universe* (Jonathan Cape, London; E.P. Dutton, New York, 1974).
—*The Next 500 Years: Life in the Coming Millennium* (Headline, London; W.H. Freeman, New York, 1995).
—*Eureka! and Other Stories: A Book of Scientific Anecdotes* (Helicon, Oxford, 1993). Also published as, respectively, *The Harrap Book of Scientific Anecdotes* (Harrap, London, 1989), and *The Book of Scientific Anecdotes* (Prometheus, Buffalo, New York, 1993).
—*Galileo and the Dolphins: Amazing but True Stories from Science* (B.T. Batsford, London, 1996).
BILSON, Elizabeth. With Yervant Terzian (eds.). *Carl Sagan's Universe* (Cornell University Press, 1998).
BOSS, Alan. *Looking for Earths: The Race to Find Solar Systems* (John Wiley, New York, 1998).
BOULTON, W.G. *The Pageant of Transport through the Ages* (London, 1931).
BOVA, Ben. With Antony R. Lewis. *Space Travel: A Writer's Guide to the Science of Interplanetary and Interstellar Travel* (Writer's Digest Books, Cincinnati, Ohio, 1997).
BRADFORD, William. *History of the Plymouth Settlement, 1608-1650*. Rendered into modern English by Valerian Paget (London, 1909).
BRIN, David. *The Transparent Society: Will Technology Force us to Choose Between Privacy and Freedom?* (Addison-Wesley, Reading, Massachusetts, 1998).

CALDER, Nigel. *Spaceships of the Mind* (British Broadcasting Corporation, 1978).

CAMERON, A.G.W. (ed.). *Interstellar Communication: A Collection of Reprints and Original Contributions* (W.A. Benjamin, New York, 1963).

CHEMICAL RUBBER COMPANY. *Handbook of Tables for Mathematics*, ed. Robert C. Weast and Samuel M. Selby (Chemical Rubber Company, Cleveland, Ohio, 1970).

—*Handbook of Chemistry and Physics*, ed. Robert C. Weast and Melvin J. Astle (CRC Press, Boca Raton, Florida, 1981).

CLARK, Grahame. *World History in New Perspective* (Cambridge University Press, 1961).

CLARKE, Arthur C. With the Editors of Time-Life Books. *Man and Space* (Time-Life International, 1970).

—*Profiles of the Future: An Inquiry into the Limits of the Possible* (Victor Gollancz, London, 1982).

—*The Promise of Space* (Hodder and Stoughton, London, 1968).

—*The Challenge of the Spaceship: Previews of Tomorrow's World* (Frederick Muller, London, 1960).

CLUTE, John. With Peter Nicholls (eds.). *The Encyclopedia of Science Fiction* (Orbit, London, 1993).

COMINS, Neil F. With William J. Kaufmann. *Discovering the Universe* (W.H. Freeman, New York, 1996).

COUPER, Heather. With Nigel Henbest. *The Guide to the Galaxy* (Cambridge University Press, 1994).

—*To the Ends of the Universe: A Voyage Through Life, Space and Time* (Dorling Kindersley, London, 1998).

COVENEY, Peter. With Roger Highfield. *The Arrow of Time: A Voyage Through Science to Solve Time's Greatest Mystery* (W.H. Allen, London, 1990).

CROSWELL, Ken. *Planet Quest: The Epic Discovery of Alien Solar Systems* (Oxford University Press, 1997).

DARLING, David. *Micromachines and Nanotechnology: The Amazing New World of the Ultrasmall* (Dillon Press, Pasippany, New Jersey, 1995).

DAUBER, Philip M. With Richard A. Muller. *The Three Big Bangs: Comet Crashes, Exploding Stars, and the Creation of the Universe* (Addison-Wesley, New York, 1997).

DAVIDSON, James Hale. With William Rees-Mogg. *The Sovereign Individual: The Coming Economic Revolution: How to Survive and Prosper in it* (Macmillan, London, 1997).

DAVIES, Paul. *Are We Alone? Implications of the Discovery of Extraterrestrial Life* (Penguin, London, 1990).

—*Other Worlds: Space, Superspace and the Quantum Universe* (J.M. Dent, London, 1978).

DAVIS, Joel. With Robert L. Forward. *Mirror Matter: Pioneering Antimatter Physics* (John Wiley, New York, 1988).

DEUTSCH, David. *The Fabric of Reality: How Much Can Our Four Deepest Theories of the World Explain?* (Penguin, London, 1997).

DIAMOND, Jared. *Guns, Germs and Steel: A Short History of Everybody for the Last 13,000 Years* (Jonathan Cape, London, 1997).

DOLE, Stephen H. *Habitable Planets for Man* (Elsevier, New York, 1970).

DORLING KINDERSLEY (computer software publishers). *Eyewitness Encyclopedia of Space and the Universe: The Ultimate Interactive Guide to Space* (DK Multimedia, London, 1996).

DRAKE, Frank. *Is Anyone Out There? The Scientific Search for Extraterrestrial Intelligence* (Souvenir Press, London, 1993).

DREXLER, K. Eric. *Engines of Creation: The Coming Era of Nanotechnology* (Fourth Estate, London, 1990).

DU CHARME, Wesley M. *Becoming Immortal: Nanotechnology, You, and the Demise of Death* (Blue Creek Ventures, Evergreen, Colorado, 1995).

DURKIN, John. *Expert Systems: Design and Development* (Prentice Hall International, London, 1994).

DYSON, Freeman. *Disturbing the Universe* (Harper and Row, New York, 1979).

—*Imagined Worlds* (Harvard University Press, 1997).

DYSON, J.E. With D.A. Williams. *The Physics of the Interstellar Medium* (Institute of Physics, Bristol, 1997).

EINSTEIN, Albert. *The Meaning of Relativity* (Chapman and Hall, London, 1967).

—With H.A. Lorentz, H. Minkowski and H. Weyl. *The Principle of Relativity: A Collection of Original Memoirs on the Special And General Theory of Relativity* (Dover Publications, New York, 1952).

EIROA, C. (ed.). *Infrared Space Interferometry: Astrophysics and the Study of Earth-Like Planets* (Kluwer, Dordrecht, 1997).
ETTINGER, Robert C.W. *The Prospect of Immortality* (Sidgwick and Jackson, London, 1965).

FEIGENBAUM, Edward. With Pamela McCorduck. *The Fifth Generation: Artificial Intelligence and Japan's Computer Challenge to the World* (Pan, London, 1984).
FEINBERG, Gerald. With Robert Shapiro. *Life Beyond Earth: The Intelligent Earthling's Guide to Life in the Universe* (Robert Morrow, New York, 1980).
FERRARA, A. With C.F. McKee, C. Heiles and P.R. Shapiro (eds.). *The Physics of the Interstellar Medium and Intergalactic Medium* (Astronomical Society of the Pacific, San Francisco, 1995).
FERRIS, Timothy (ed.). *The World Treasury of Physics, Astronomy and Mathematics* (Little, Brown, Boston, 1991).
—*Coming of Age in the Milky Way* (William Morrow, New York, 1988).
FINNEY, Ben R. With Eric M. Jones (eds.). *Interstellar Migration and the Human Experience* (University of California Press, Los Angeles, 1985).
FORWARD, Robert L. *Indistinguishable from Magic: Speculations and Visions of the Future* (Baen, Riverdale, New York, 1995).
FRASER, Gordon. With Egil Lillestøl and Inge Sellevåg. *The Search for Infinity: Solving the Mysteries of the Universe* (Mitchell Beazley, London, 1994).

GAMBLE, Clive. *Timewalkers: The Prehistory of Global Civilisation* (Harvard University Press, 1994).
GAMOW, George. *One . . . Two . . . Three . . . Infinity* (New York, 1947).
GARDNER, Martin. *Relativity for the Million* (Macmillan, New York, 1962).*
GILLETT, Stephen L. *World-Building: A Writer's Guide to Constructing Star Systems and Life-Supporting Planets* (Writer's

* An updated version of Gardner's book, retitled *Relativity Simply Explained*, was published by Dover Publications, Mineola, New York, in 1997. It remains, in my opinion, by far the easiest to understand book on relativity yet written.

Digest Books, Cincinnati, Ohio, 1996).
GOLDSMITH, Donald. *Worlds Unnumbered: The Search for Extrasolar Planets* (University Science Books, Sausalito, California, 1997).
GOODWIN, Simon. With John Gribbin. *Empire of the Sun: Planets and Moons of the Solar System* (Constable, London, 1998).
GRUN, Bernard. *The Timetables of History: A Chronology of World Events from 5000 BC to the Present Day* (Thames and Hudson, London, 1975).

HARDY, David. With Patrick Moore. *The New Challenge of the Stars* (Mitchell Beazley, London, 1977).
HARRIS, Philip Robert. *Living and Working in Space* (John Wiley, Chichester, 1996).
HART, Michael. With Ben Zuckerman (eds.). *Extraterrestrials: Where are They?* (Cambridge University Press, 1995).
HARTMANN, William K. With Ron Miller. *The Grand Tour: A Traveller's Guide to the Solar System* (Workman, New York, 1993).
HAWKING, Stephen. With Martin J. Rees. *Before the Beginning: Our Universe and Others* (Addison-Wesley, New York, 1997).
HEIDMANN, Jean. *Extraterrestrial Intelligence* (Cambridge University Press, 1995).

JAFFE, Bernard. *Michelson and the Speed of Light* (Heinemann, London, 1960).
JAKOSKY, Bruce. *The Search for Life on Other Planets* (University of Colorado Press, 1998).
JASTROW, Robert. *Until the Sun Dies* (Souvenir Press, London, 1977).
—*Journey to the Stars: Space Exploration – Tomorrow and Beyond* (Bantam Books, New York, 1989).
JENKINS, Adrian. *The Number File* (Tarquin Publications, Stradbroke, Norfolk, 1985).

KAHN, David. *The Codebreakers: The Story of Secret Writing* (Weidenfeld and Nicolson, London, 1966).
KAKU, Michio. *Visions: How Science will Revolutionize the 21st Century* (Doubleday, New York, 1997).
—*Hyperspace: A Scientific Odyssey Through Parallel Universes,*

Time Warps, and the Tenth Dimension (Oxford University Press, 1995).

KAUFMANN, William J. *Relativity and Cosmology* (Harper and Row, New York, 1977).

KRAUSS, Lawrence M. *The Physics of Star Trek* (Basic Books, New York, 1995).

—*Beyond Star Trek: Physics from Alien Invasions to the End of Time* (Basic Books, New York, 1997).

LALLEMENT, R. With R. von Steiger and M.A. Lee (eds.). *The Heliosphere in the Local Interstellar Medium* (Kluwer, Dordrecht, 1996).

LAPP, Ralph E. With the Editors of LIFE. *Matter* (Time-Life International, New York, 1963).

LEMONICK, Michael D. *Other Worlds: The Search for Life in the Universe* (Simon and Schuster, New York, 1998).

LEWIS, John S. *Mining the Sky: Untold Riches from the Asteroids, Comets, and Planets* (Addison-Wesley, Reading, Massachusetts, 1996).

MALIN, David. *A View of the Universe* (Cambridge University Press, UK; Sky Publishing, Cambridge, Massachusetts, 1995).

MALLOVE, Eugene. With Gregory Matloff. *The Starflight Handbook: A Pioneer's Guide to Interstellar Travel* (John Wiley, New York, 1989).

MANCHESTER, William. *A World Lit Only by Fire: The Medieval Mind and the Renaissance: Portrait of an Age* (Little, Brown, Boston, 1993).

MARSHAK, R.E. (ed.). *Perspectives in Modern Physics* (Interscience Publishers, New York, 1966).

MARX, Leo. With Merritt Roe Smith (eds.). *Does Technology Drive History? The Dilemma of Technological Determinism* (Massachusetts Institute of Technology Press, 1994).

MAULDIN, John H. *Prospects for Interstellar Travel* (American Astronautical Society, San Diego, 1992).

MILLMAN, Gregory J. *The Vandals' Crown: How Rebel Currency Traders Overthrew the World's Central Banks* (The Free Press, New York, 1995).

MISNER, Charles W. With Kip S. Thorne and John Archibald Wheeler. *Gravitation* (W.H. Freeman, San Francisco, 1973).

MOORE, Patrick. *Atlas of the Universe* (Philip's, Reed Consumer Books, London, 1994).

MORRISON, Philip and Phylis. *Powers of Ten: About the Relative Size of Things in the Universe and the Effects of Adding Another Zero* (Scientific American Books, New York, 1962).

MOSZKOWSKI, Alexander. *Conversations with Einstein* (Sidgwick and Jackson, London, 1972).

NAYLOR, Chris. *Build Your Own Expert System: Artificial Intelligence for the Aspiring Microcomputer* (Sigma Technical Press, Wilmslow, Cheshire, 1987).

NEON (a pseudonym for Marion W. Acworth). *The Great Delusion: A Study of Aircraft in Peace and War* (London, 1927).

NICOLSON, Iain. *The Road to the Stars* (Westbridge Books, Newton Abbot, Devon, 1978).

NUMELIN, Ragnar. *The Wandering Spirit: A Study of Human Migration* (London, 1937).

O'NEILL, Gerard K. *The High Frontier: Human Colonies in Space* (Jonathan Cape, London, 1977).

PAGE, Lou Williams. With Thornton Page (eds.). *Starlight: What It Tells Us about the Stars* (Vol. 5 of the Sky and Telescope Library of Astronomy) (Macmillan, New York, 1967).

PATERSON, Antoinette Mann. *The Infinite Worlds of Giordano Bruno* (Thomas, Springfield, Illinois, 1970).

PECKER, Jean-Claude. *The Future of the Sun* (McGraw-Hill, New York, 1992).

PELLEGRINO, Charles. *Flying to Valhalla* (William Morrow, New York, 1993).*

PERRYMAN, A.C. With the Hipparcos Science Team. *The Hipparcos and Tycho Catalogues* (of stellar positions) (European Space Agency, Noordwijk, Netherlands, 1997).

—With R.W. Sinnott. *Millennium Star Atlas* (Sky Publishing, Cambridge, Massachusetts; the European Space Agency, Noordwijk, 1997).

* A work of fiction, but with a most valuable Appendix on the science of interstellar flight.

READER'S DIGEST ASSOCIATION. *The Last Two Million Years* (London, 1973).
REES, Martin. With Remo Ruffini and John Archibald Wheeler. *Black Holes, Gravitational Waves and Cosmology: An Introduction to Current Research* (Gordon and Breach, New York, 1974).
REGIS, Ed. *Nano: The True Story of Nanotechnology – The Astonishing New Science that will Transform the World* (Bantam, London, 1997).
RESTON, James. *Galileo: A Life* (Cassell, London, 1994).
RONAN, Colin. *The Universe: From the Big Bang to the End of Time* (Marshall Publishing, London, 1998).
ROWAN-ROBINSON, Michael. *Our Universe: An Armchair Guide* (Longman, London, 1990).
RUELLE, David. *Chance and Chaos* (Princeton University Press, 1992).

SAGAN, Carl. *Pale Blue Dot: A Vision of the Human Future in Space* (Random House, New York, 1994).
—With I.S. Shklovskii. *Intelligent Life in the Universe* (Holden-Day, San Francisco, 1966).
SANGER, Eugen. *Space Flight: Countdown for the Future* (McGraw-Hill, New York, 1965).
SAVAGE, Marshall T. *The Millennial Project: Colonising the Galaxy in Eight Easy Steps* (Little, Brown, Boston, 1994).
SCHECHNER GENUTH, Sara. *Comets, Popular Culture, and the Birth of Modern Cosmology* (Princeton University Press, 1997).
SCHMIDT, Stanley. With Robert Zubrin (eds.). *Islands in the Sky: Bold New Ideas for Colonizing Space* (John Wiley, New York, 1996).
SCHNEIER, Bruce. *Applied Cryptography: Protocols, Algorithms, and Source Code in C* (John Wiley, New York, 1994).
SHOSTAK, G. Seth (ed.). *Progress in the Search for Extraterrestrial Life* (Astronomical Society of the Pacific, San Francisco, 1995).
SINGER, Dorothea Waley. *Giordano Bruno: His Life and Thought* (Henry Schuman, New York, 1950).
STANISLAW, Joseph. With Daniel Yergin. *The Commanding Heights: The Battle between Government and the Marketplace that is Remaking the Modern World* (Simon and Schuster, New York, 1998).

STORK, David G. (ed.). *HAL's Legacy: 2001's Computer as Dream and Reality* (The MIT Press, Cambridge, Massachusetts, 1997).
STRONG, James. *Flight to the Stars: An Enquiry into the Feasibility of Interstellar Flight* (Temple Press Books, London, 1965).

THORNE, Kip S. *Black Holes and Time Warps: Einstein's Outrageous Legacy* (W.W. Norton, New York, 1994).
TOVMASYAN, G.M. (ed.). *Extraterrestrial Civilisations* (Academy of Sciences of the Armenian SSR; Proceedings of the First All-Union Conference on Extraterrestrial Civilisations and Interstellar Communications, Israel Programme for Scientific Translations, Jerusalem, 1967).

UPGREN, Arthur. *Night Has a Thousand Eyes: A Naked Eye Guide to the Sky, Its Science, and Lore* (Plenum Press, New York, 1998).

VISSER, Matt. *Lorentzian Wormholes: From Einstein to Hawking* (Oxford University Press, 1996).

WEATHERFORD, Jack. *Savages and Civilisation: Who Will Survive?* (Fawcett Columbine, New York, 1994).
WHEELER, J.A. *Geometrodynamics* (Academic Press, New York, 1962).
WHITROW, G.J. *What is Time?* (Thames and Hudson, London, 1972).
WILLIAMS, L. Pearce (ed.). *Relativity Theory: Its Origins and Impact on Modern Thought* (John Wiley, New York, 1968).
WILLIAMS, Trevor I. *The Triumph of Invention: A History of Man's Technological Genius* (Macdonald Orbis, London, 1987).
WOLF, Fred Alan. *Parallel Universes: The Search for Other Worlds* (Touchstone, New York, 1988).*

ZUBRIN, Robert. With Richard Wagner. *The Case for Mars: The Plan to Settle the Red Planet and Why We Must* (The Free Press, New York, 1996).

---

* The subtitle of this book refers to other universes, not alien planets.

Index

Note: The letter *n* added to a page number indicates a footnote, *g* a glossary entry, and a small number an endnote. Where the same endnote number occurs more than once on a page its position is given in brackets.

Index